Ecology of Plant-Derived Smoke

Ecology of Plant-Derived Smoke

ITS USE IN SEED GERMINATION

Lara Vanessa Jefferson, Marcello Pennacchio, and Kayri Havens

ILLUSTRATIONS BY DAVID S. SOLLENBERGER

OXFORD
UNIVERSITY PRESS

OXFORD
UNIVERSITY PRESS

Oxford University Press is a department of the University of Oxford.
It furthers the University's objective of excellence in research, scholarship,
and education by publishing worldwide.

Oxford New York
Auckland Cape Town Dar es Salaam Hong Kong Karachi
Kuala Lumpur Madrid Melbourne Mexico City Nairobi
New Delhi Shanghai Taipei Toronto

With offices in
Argentina Austria Brazil Chile Czech Republic France Greece
Guatemala Hungary Italy Japan Poland Portugal Singapore
South Korea Switzerland Thailand Turkey Ukraine Vietnam

Oxford is a registered trade mark of Oxford University Press
in the UK and certain other countries.

Published in the United States of America by
Oxford University Press
198 Madison Avenue, New York, NY 10016

© Oxford University Press 2014

Library of Congress Cataloging-in-Publication Data
Jefferson, Lara Vanessa.
   Ecology of Plant-Derived Smoke : its use in seed germination / Lara Vanessa Jefferson,
Marcello Pennacchio, and Kayri Havens; illustrations by David S. Sollenberger.
      p.   cm.
   Includes bibliographical references and index.
   ISBN 978-0-19-975593-6 (alk. paper)
   1. Germination. 2. Plants—Effect of smoke on. 3. Smoke—Physiological effect. I. Pennacchio,
Marcello. II. Havens, Kayri. III. Title.
   QK740.J34 2014
   571.8'62—dc23
                              2013026073
9780199755936

9 8 7 6 5 4 3 2 1

Printed in the United States of America on acid-free paper

To my wonderful mother, Brenda Doreen Jefferson (LVJ).

To my late nephew, Michael Gismondi, and to my two sons, Ryan Dharius Pennacchio and Aiden Michael Pennacchio (MP).

For my husband Eric, and his daughter, Zoe Young (KH).

CONTENTS

PREFACE

Ever since our early ancestors learned how to make fire, the smoke it produces has also found a number of uses. Plants that release chemical substances when burned have served as medicines, incense for magico-religious ceremonies, recreational drugs, perfumes, and for flavoring food and beverages. These and other uses have, in the past and present, spawned large organizations that have earned billions of dollars from the sale of smoke products. One needs look only at tobacco or the Arabian incense trade that existed at the time of Christ to realize the enormous socioeconomic importance of smoke. The earnings from these two industries alone have rivaled those of the contemporary oil industry.

Plant-derived smoke has also been employed as a variety of lucky charms. Native Americans once used it to lure game and even lovers. Elsewhere in Africa and Europe, it had a more interesting use, one that is of greater relevance to the smoke ecology story. Members of Zulu tribes, for example, used smoke as a fertility charm to promote better crops, a lesson they may have learned by observing nature. Those tribes that used smoke for this purpose were almost certainly unaware that chemicals in the smoke may have been responsible for improving their crops. It is only in more recent times that researchers have discovered the active constituent in smoke, karrikinolide, as well as other related substances. These substances work both synergistically and antagonistically, perhaps ensuring that conditions are not only right for germination, but also for the continued growth and improved vigor of plant communities.

To date, most smoke-germination studies have focused primarily on plants occurring in the fire-prone Mediterranean environments of the western United States, Western Australia, and South Africa. These environments are among the richest floristic regions of the world, with thousands of species occurring in each of them. Not all plant species respond to smoke, however. Some require other environmental cues, while many require multiple cues to germinate. The dormancy mechanism existing in each species often determines which cue or groups of cues are required. A brief account of all relevant types, along with a brief history of smoke ecology and evolution, inhibition of germination by smoke, and the chemistry of smoke products, is therefore provided. Also described are the probable mechanisms of action for these chemicals, the effect of smoke and its products on plant growth and vigor, the different smoke application types, and implications for conservation and land management. Finally, a total of 1,355 plant species, whose seed have been tested for their response to smoke and its products are listed, along with distribution and other relevant information.

This book should serve as an excellent springboard for scientists wanting to learn about the fascinating field of smoke-related seed germination, ecology, evolutionary, and physiology. The specific purposes of this book are to provide a comprehensive literature review of research conducted to date on the subject. The intended audiences are: (1) the scientific community in general, specifically those interested in the ecology, evolution, or physiology of plants; (2) land managers who can use this knowledge in their practices; and (3) biology or science students who will benefit from the history and background of this science and the interdisciplinary approach necessary for studies of this type.

ACKNOWLEDGMENTS

A book such as this cannot be written without the generous help and support of many people, all of whom we would like to sincerely thank. Thanks to Leora Siegel, director of the Lenhardt Library of the Chicago Botanic Garden and her volunteers for their assistance. We would especially like to acknowledge the help of Stacy Stoldt, also of the Lenhardt Library, and Christine Schmid, for tirelessly tracking down articles and books for us. Libraries throughout Illinois, and elsewhere in the United States, assisted greatly with interlibrary loans. The editors and staff at Oxford University Press are thanked for making the process of publishing this book a smooth and enjoyable one, always offering friendly and useful advice.

We would also like to acknowledge the enormous contribution that Carol Line, executive director of Fernwood Botanical Garden and Nature Preserve in Michigan, United States, and Samuel J. Goldman, a volunteer at the Chicago Botanic Garden, have made to this project. Bob Meyer, also a volunteer at Chicago Botanic Garden, was instrumental in helping us with the illustrations. The following people are similarly thanked for reading parts or all of the manuscript, or assisting with other aspects of preparing it: Judy Cashen (manager of Volunteer Services), Carol Baskin, Jerry Baskin, Susanne Masi, Gail Kushino, Luisa Miller, Boyce Tankersley, Monica Vachlon, Marley and Michael Sackheim, Randy Hetzel, and Lori Sollenberger, as well as our families and friends. We would especially like to thank Brenda and Roger Jefferson for their encouragement and for babysitting. Finally, we would like to acknowledge the wonderful and generous support of the staff and members of the Plant Conservation Alliance, especially Ms. Margaret Peggy Olwell.

Ecology of Plant-Derived Smoke

Introduction

Since the appearance of plant life on earth, approximately 350 million years ago, a variety of extreme forces have contributed to its enormous diversity. Climatic fluctuations, fires, massive volcanic eruptions, asteroids, and other impacts have all been responsible, in one way or another, for changes to past and present plant species and communities around the world. As a result, there are now more than 250,000 recognized species of plants with many more likely to be discovered in the future. This figure does not, however, include those species that are now extinct. Unless conservation efforts are improved, others will similarly disappear.

Early humans may have also played a contributing role in diversifying the flora, especially through their use of fire (Pyne 2001; Vogl 1974). It is not clear when humans commenced using this tool, but is thought to have started approximately 1.6 million years ago (Kempe 1988). Since then, fire has followed our ancestors out of Africa and across the globe. The charcoal left in the wake of those early human diasporas "has become the spoor of early hominids," according to the world's leading authority on the history of fire, Stephen J. Pyne of Arizona State University. In fact, their very travels can be retraced as a result of its use and long-term preservation.

While our ancestors used fire for a variety of purposes, it is unlikely they used it to enhance or promote crops until well after regular farming practices commenced approximately 10,000 years ago. Specifically, the use of smoke was first considered a fertility charm rather than a tool. Early records suggest that peasant farmers living in the vicinity of the Eifel Mountains of Germany and people of Great Britain's Isle of Man believed that smoke wafting over their crops was an omen for bountiful harvests (Frazer 1922; Pennacchio et al. 2007a, 2010). Members of the Matabeles sect (or Ndebele tribesmen) of southern Africa's Zulu people held similar beliefs and burned several species of plants as part of this practice, including *Myrica serrata* (Hutchings et al. 1996), *Euphorbia triangularis* (Pooley

1

1993), *Ipomoea crassipes*, and *I. pellita* (Gerstner 1939). These were lessons probably learned by observing and then copying nature in action, a concept known as biomimicry.

Evidence for deliberately treating seeds with heat and smoke prior to sowing them was reported as early as 1632 by Gabriel Sagard, a French Récollet missionary. In 1624 Sagard, like so many of his Franciscan brothers, was sent to study the customs and mores of the tribes of New France. This was an area of North America originally colonized by the French that extended from Newfoundland to the Gulf of Mexico. During his time with the tribes of the Huron people, Sagard recorded they suspended special germination boxes, lined with multiple layers of soil and pumpkin seeds, above fires where the smoke plumes formed. This, he claimed, increased the number of sprouters.

Modi (2002, 2004) reported that similar smoke-over-fire storage methods currently used by native subsistence farmers in South Africa improved not only seed germination but also seed quality and seedling vigor in at least two land races of maize. The Mam Maya people of the highlands in Guatemala, in contrast, used copal resins from *Bursera* trees (Burseraceae), which are related to those that produce frankincense and myrrh, to improve germination of maize seeds (*Zea mays*). In a ritual called "pomixi," or "copal of maize," resins from the bark were combined with a drop of sacrificial blood and burned to fumigate the seed (Wagley 1957). It is doubtful, though, that any of these people were aware that substances in the smoke were responsible for the activities they were observing.

Perhaps the first scientific record linking smoke with its ability to promote seed germination was reported by Wicklow in 1977. Wicklow showed that the seeds of the chaparral species, *Emmenanthe penduliflora*, germinated when treated with burned plant stem segments. The ash of incinerated stem segments, in contrast, had no effect. Wicklow concluded that products generated during fire had somehow triggered germination by "providing a germination cue in the form of burned plant remains."

The first real use of smoke as a germination cue did not occur until 1990, when the team of De Lange and Boucher reported that germination of the seeds of *Audouinia capitata*, a representative of the Western Cape fynbos, was significantly promoted by smoke. The Cape Floristic Region of South Africa is a biodiversity hotspot with over 9,600 plant species (Conservation International 2013). Since then, other species with smoke-responsive seeds have been discovered (Brown 1993a,b; Brown et al. 1994; Brown et al. 2003; Pierce et al. 1995), including from the Southwest Botanical Province of Western Australia (Dixon et al. 1995; Roche et al. 1997a,b), and the chaparrals of the Californian Floristic Province (Fotheringham et al. 1995; Keeley and Fotheringham 1998a). These too are biodiversity hotspots (Conservation International 2013).

For the most part, these studies have focused on the seeds of plants that commonly occur in fire-prone ecosystems (over 80% of species listed in this compendium). It is within these sorts of environments that the response of seeds to natural cues, such as smoke, might be correlated with life history traits. Growth form, plant family, seed mass and size, or dispersal mode, in contrast, were not as useful for predicting whether or not a species was likely to respond to the effects of smoke (Brown et al. 2003; Dixon et al. 1995). Brown et al. (2003) suggested, instead, that the regeneration strategy of South African fynbos species was more ideal for predicting smoke responsiveness. Serotinous species (canopy-stored seed banks), and those that resprout following a fire, are less prone to set smoke-responsive seeds. In contrast, nonserotinous species (soil-stored seed banks), and those that are solely dependent on seeds for regeneration, that is, obligate seeder species, are more likely to respond to smoke.

Pate et al. (1990) compared congeneric pairs of sprouting and nonsprouting species in southwestern Australian heathlands and suggested that those species that resprout following fires typically exhibited slower growth rates, higher concentrations of starch in the roots, lower root:shoot ratios, and delayed flowering. Similar results have since been reported for epacrids (Bell et al. 1996) and Bossiaea species (Hansen et al. 1991). Those species that re-sprout after fires tend to allocate more energy into survival, while the nonsprouters that do not survive fire allocate more energy into fast seedling growth and reproduction (refer to Bond and Wilgen 1996 for a more comprehensive review). It would seem plausible then that the seeds of the nonsprouters have evolved to germinate upon contact with smoke when environmental conditions are conducive to growth and survival.

Plants of other Mediterranean regions also produce seeds that respond to smoke (Pérez-Fernández and Rodríguez-Echeverría 2003; Crosti et al. 2006; Gómez-González et al. 2011; Çatav et al. 2012), as do semiarid (Anderson et al. 2012; Schwilk and Zavala 2012), arid (Pierce et al. 1995), temperate (Keith 1997; Pennacchio et al. 2005; Jefferson et al. 2007; Tsuyuzaki and Miyoshi 2009), subtropical (Tang et al. 2003; Zuloaga-Aguilar et al. 2011), and tropical (Singh and Raizada 2010; Kandari et al. 2011) plants. Interestingly, wet tropical ecosystems that generally experience lower frequencies of fire in comparison to their temperate and Mediterranean counterparts also include smoke-responsive species.

In addition, plants of the granite outcrops of sclerophyllous woodlands of Western Australia tend to be fire responsive, even though their habitat does not promote fire. For example, germination of seeds of *Verticordia endlicheriana* var. *angustifolia*, *V. fimbrilepis* ssp. *australis*, and *V. staminosa* ssp. *staminosa* (Myrtaceae) were promoted by smoke (Cochrane et al. 2002). Despite granite outcrops not being prone to fires, they are often exposed to the smoke of surrounding forest fires, which could perhaps explain why the plants that grow on them are smoke responsive. It is possible also that their ancestors originally inhabited fire-prone environments. Many species that no longer occur in ecosystems that burn frequently have responded to smoke and its products. These species include cultivated lettuce seeds (*Lactuca sativa*; Drewes et al. 1995; Light et al. 2002), celery (*Apium graveolens*; Thomas and van Staden 1995), red rice (Doherty and Cohn 2000), and wild oats (Adkins and Peters 2001). There is always the possibility that there are several different germination stimulants in smoke, acting through different mechanisms.

Smoke contains thousands of different compounds (Maga 1988). Many of them are not exclusive to smoke and commonly occur in soil and air and promote germination in some cases. They include nitrogen oxides and nitric oxides (Keeley and Fotheringham 1997), ethylene (Adkins and Ross 1981), and carbon dioxide (Bassi et al. 1975), to name a few. Germination of *Emmenanthe penduliflora* (Haemodoraceae) seeds significantly increased, for example, in response to both nitrogen oxides (Keeley and Fotheringham 1997) and smoke water. In other cases, the seeds of some species responded to smoke treatments, but not to pure nitrogen oxides (Keeley and Fotheringham 1998a,b), or ethylene (van Staden et al. 1995). It was thought that other compounds in the smoke were responsible for promoting germination.

Chemicals That Affect Germination

The active constituents of smoke have eluded scientists until recently. Keeley and Pizzorno (1986) first attempted to identify germination stimulants from charred wood. Several hemicelluloses, sugars, polyethylene glycol, acids, and acid hydrolyzed woods were identified, but were not the much sought after active compounds. Nevertheless, they drew the following conclusions:

The stimulatory effect of smoke is (1) produced by a wide variety of wood types, (2) not dependent on light, (3) produced by low amounts of charred material on a weight to volume basis, (4) water soluble, (5) active after 24-hour exposure, and (6) produced by heating at 175°C for 30 minutes.

Baldwin et al. (1994) also attempted to isolate and identify the active compound from smoke water using a bioassay-driven fractionation process. Seventy-one compounds were isolated, but none of them stimulated germination to the same degree as the crude smoke extract. Searches by van Staden et al. (1995) and Jäger et al. (1996) were likewise unsuccessful, but did make several important discoveries, for example, that the active substance was from common plant products and that it was water soluble, thermostable, and active in low concentrations.

The discovery of an active substance was not made until 2004, when it was simultaneously isolated by the Western Australian team of Flematti et al. (2004) and the South African team of van Staden et al. (2004). Working independently, these two groups showed that a water-soluble combustion product of cellulose, belonging to the butenolide class of compounds, was responsible for promoting germination in lettuce seeds. The butenolides are natural substances derived from various sources, including the filamentous fungi genus, Fusarium (Wang et al. 2006). They exhibit a variety of biological activities, such as promoting and inhibiting seed germination (see List of Plants), inhibiting shoot branching (Chen et al. 2009), inducing hyphal branching in arbuscular mycorrhizal fungi (Chen et al. 2009), toxicity (Burmeister et al. 1971), and antibiosis (Berrie et al. 1967). The active butenolide substance in smoke has recently been named karrikinolide (KAR$_1$; Commander et al. 2008), a name derived from the Australian Nyungar

Aboriginal word for smoke, *karrick*. The Aboriginal people have traditionally used plant-derived smoke for a variety of purposes (see Pennacchio et al. 2010). Not all species respond to KAR₁, however. Downes et al. (2010) recently showed that the fire ephemeral species, *Tersonia cyathiflora*, when tested with KAR₁, did not germinate, but did so in response to smoke water. They concluded other substances in the smoke water preparation must have stimulated the germination. Long et al. (2011) suggested that the effects of KAR₁ are dependent on many factors, including light, temperature, and dormancy state of the seeds. Species cannot, therefore, simply be classed as KAR₁ responsive or nonresponsive.

Since the discovery of karrikinolide (3-methyl-2*H*-Furo[2,3-*c*]pyran-2-one; Fig. 1A), its biosynthesis has been described by several researchers (Flematti et al. 2007; Nagase et al. 2008; Sun et al. 2008). This compound promotes germination in a variety of plant species, irrespective of whether or not they are from fire-prone habitats (for the complete list, refer to our List of Plants). Karrikinolide, which has recently been synthesized artificially from pyromeconic acid (Flematti et al. 2005), and whose bioactivity resembles that identified in other structurally related butenolides (see Pepperman and Bradow 1988; Pepperman and Cutler 1991), is active in concentrations as low as 10⁻⁹ M in germination of Grand Rapids lettuce seeds. A number of synthetic karrikinolide analogs (Sun et al. 2008) have also emerged and have likewise been shown to promote germination in the seeds of various plants, including *Lactuca sativa*, *Emmenanthe penduliflora* and *Solanum orbiculatum* (Flematti et al. 2007), as well as five analogous substances that have been named karrikins (Flematti et al. 2009). One other substance that may be of interest is the cyanohydrin glyceronitrile. In the presence of water, glyceronitrile is slowly hydrolyzed into cyanide, which stimulates germination in various fire-responsive plant species (Flemmatti et al. 2011a) but not others (Downes et al. 2013).

Of equal importance to the smoke ecology story is the ability for smoke to inhibit seed germination (Adkins and Peters 2001; Brown et al. 1993; Drewes et al. 1995; Jefferson et al. 2007; Light et al. 2002; Merritt et al. 2006; Pennacchio et al. 2005, 2007b; Tsuyuzaki and Miyoshi 2009). Inhibitors in smoke are often present with the promoters, but they may require longer exposure times to smoke to mediate their effects (Pennacchio et al. 2005, 2007b). Furthermore, they can be rinsed off with water (Light et al. 2002). The effects of the promoters, in contrast, appear to be irreversible once induced (Baldwin et al. 1994, Pennacchio et al. 2007b).

An interesting inhibitory substance, also belonging to the butenolide class of compounds, was recently discovered among the thousands of substances in smoke. Light et al. (2010) reported that 3,4,5-trimethylfuran-2(5*H*)-one (Fig. 1B) not only inhibited germination in seeds but also diminished the effects of the active promotive substance. Interestingly, an antibiotic called patulin (Fig. 1C), which is itself a butenolide and resembles karrikinolide in structure, has been shown to inhibit germination in lettuce seeds (Berrie et al. 1967). Plants and other organisms often produce a variety of substances, some of which act synergistically while others are antagonistic.

The promotive and inhibitory substances in smoke appear to exert dual regulation in seed germination. This has been reported in various species, including *Arabidopsis thaliana* (Pennacchio et al. 2007b). The presence, abundance and effects of these compounds may depend on many factors, such as the type of organic matter burned, that is, species of plants burned and the metabolites they produce, as well as the level of exposure to the compounds (Pennacchio et al. 2007b). Although the reason why smoke-derived inhibitors exist is not yet clear, one hypothesis is that they may be a mechanism that prevents seed germination from occurring until there is sufficient rainfall for the species to grow and establish. This is often considered the role dormancy plays in seed biology.

Multiple Cues

Seed dormancy is thought to allow germination to occur only when environmental conditions are favorable for the growth and survival of plants (i.e., optimize reproduction). While the effects of fire on mature vegetation tend to provide ideal environmental conditions, such as eliminating competition, providing more space, increasing light and soil temperature from direct sunlight, creating higher nitrogen levels, and destroying allelopathic compounds, it does not guarantee that there will be sufficient water for seedling survival and growth.

A variety of dormancy classes have been described by renowned seed biologists, Carol and Jerry Baskin, of the University of Kentucky (Baskin and Baskin 1998). Their system of classification includes five classes of dormancy based on the original works of Nikolaeva (1969). These include morphological dormancy (MD), morphophysiological dormancy (MPD), physiological dormancy (PD), physical dormancy (PY), and combinational dormancy (PY + PD).

The kind of seed dormancy present in any given species will often determine the type of environmental cue required for germination to occur. Two or more cues may be needed to promote it (Dixon et al. 1995; Keith 1997; Thomas et al. 2003; Tieu et al. 2001a). The list of candidates include gibberellic acid (Cochrane et al. 2002; Thomas and Davies 2002; van Staden et al. 1995), cold stratification (Jefferson et al. 2007), warm stratification (Merritt et al. 2007), scarification (Brown and Botha 2004; Morris 2000), soil storage (Baker et al. 2005; Keeley and Fotheringham 1998a; Tieu et al. 2001b), and heat shock (Keeley 1987; Keith 1997; Tieu et al. 2001a). These occur as synergistic or unitive responses.

Synergistic (Thomas et al. 2003) responses usually occur when one cue promotes germination, but is significantly improved when treated with a second or third cue. For example, a combination of heat treatments followed by a subsequent treatment with smoke had a synergistic effect on the germination of the southwestern Australian species, *Actinotus leucocephalus*, *Anigozanthus manglesii*, *Loxocarya striatus*, *Sowerbaea laxiflora*, and *Stylidium affina* (Tieu et al. 2001a; refer also to Keith 1997). Unitive responses are different in that they occur only when there is an increase in germination from two or more simultaneous cues. Germination is not improved when treated with just one or any of the other cues. This was demonstrated in two *Baeckia* species (Thomas et al. 2003; refer also to Gilmour et al. 2000; Kenny 2000; Morris 2000; Hidayati et al. 2012).

Several environmental cues will naturally also prevent germination in some species, but can be counteracted with smoke, especially in high temperatures (Jain et al. 2006; Thomas et al. 2007). Thomas et al. (2007) observed this for several Ericaceae and Myrtaceae species at 25, 50, and 75°C. Jain et al. (2006) suggested that smoke promoted germination of tomato seed at 10 and 40°C. It is possible that higher than usual temperatures may denature germination inhibitors or change the substrate structure, paving the way for the promotive substances to mediate their effects.

Multiple environmental cues may provide signals intricately linked with physiological feedback loops within the seeds. They ensure that not all seeds germinate at once or are triggered by only one cue. This could explain why smoke does not trigger a mass germination response in any given area even though very low concentrations of karrikinolide are required to promote germination in some species. The role that smoke plays in the germination process requires an understanding of the positive and negative feedback loops that occur between external and internal factors. The variation in responses from one species to the next contributes to the biodiversity of plant communities and is what makes this area of research so fascinating.

While variation in response from one species to the next is very evident, variation in response from one study to the next of the same species is now commonly reported. Such variation within a species may be the result of a number of factors relating to characteristics of the seed batch, including geographic location of the source, age of the seeds, and susceptibility to transience in dormancy state, to name a few. Seed characteristics, treatments (i.e., some species will require multiple cues), germination conditions (Table 1), and combinations thereof, need to be considered when planning a study of the types listed here, when publishing results, and when analyzing and making comparisons with other studies. The differences in the approaches adopted by many researchers may account for the variation in responses from one study to the next, and reported for many of the species listed in this compendium.

Evolution

The different classes of dormancy (described above) were overlaid by Baskin and Baskin (2004) onto a family tree of seed phylogeny developed earlier by Martin (1946). Martin (1946) used morphological traits, including embryo size and position within the seeds, to suggest possible origins and evolutionary relationships. Where dormancy is concerned, both PD and MD are the most widespread of the

TABLE 1

Factors to be considered when testing and analyzing the effects of smoke on seed germination

Seed Characteristics	Treatments (singular or in combination)	Germination Conditions
Source of seeds	Smoke treatment, including	Temperature
Age of seeds	1. Smoke type:	Moisture
Embryo maturity	¤ Aerosol smoke	Light/dark
Storage conditions and period	¤ Aqueous smoke	Humidity
Dormancy type	¤ Karrikinolide	Germination media
Viability	¤ Karrikins	Osmotic potential
	¤ Glyceronitrile	Water purity
	¤ Cyanide	Sterility
	¤ Charate	(a) Ex situ versus in situ
	¤ Ash	trials
	¤ Nitrogen oxide	Pseudoreplication in
	¤ Strigolactones	experimental design
	¤ Synthetic strigol analog, GR24	
	2. Smoke dose or concentration	
	3. Smoke application period	
	Heat shock, including	
	1. Temperature	
	2. Application period	
	Stratification, including	
	1. Wet/dry	
	2. Cold/warm	
	3. Application period	
	Scarification, including	
	1. Degree of seed coat removal	
	Burial in soil, including	
	1. Season	
	2. Period	
	3. Soil type	
	Gibberellic acid, including	
	1. Dose	
	2. Application period	
	Materials used to generate the smoke (some may contain substances that are naturally inhibitory)	

five classes (i.e., spread over the entire tree of seed phylogeny). The climate present during the time since seed-producing plants first appeared in the fossil record (Cretaceous Period: 146–64.5 million years ago) gives some indication of how and when dormancy developed (Baskin and Baskin 1998) and why PD is so widespread. Areas that experienced high rainfall since the Cretaceous Period include tropical rainforests with a high number of species whose seeds are nondormant. Regions that have experienced low rainfall (e.g., southcentral Africa, westcentral South America and central Asia) or variable rainfall (e.g., western and central Australia, and central North America) tend to produce species that commonly exhibit seed dormancy. Not only were these seed dormancy classes shaped by rainfall and temperature but probably by fire as well. It is tempting to suggest, therefore, that the PD mechanism broken by smoke application in smoke-responsive species evolved later as a result of fire during the Pliocene Epoch (5.3–1.8 million years ago) of the Tertiary Period. It is during this period that the climate became drier and harsher and, thus, germination in the presence of adult plants would have been detrimental to many individual plants.

In reviewing the effects of fire on grasslands, Vogl (1974) suggested that fire was a natural selective force in the resultant biodiversity of grassland species, and that the heat those fires generated had a mutagenic effect. Those plants that were still in flower, but were not destroyed by the fire, were often undergoing meiosis-gametogenesis, that is, production of gametes (pollen or ovules), which could be affected by the heat from the fire. Vogl (1974) suggested that the heat likely induced genetic mutations (changes in the DNA sequence), upon which natural selection could act.

Recent evolutionary research has reported on the role of heat shock mutations in the evolution of plants. Waters and Schaal (1996) reported that heat shock (40°C for 2 hours) induced changes in the DNA of leaves of 2-week old *Brassica nigra* seedlings and that those changes were passed on to the following F1 generation. Heat shock is known to alter the patterns of gene expression and protein translation by producing heat shock proteins (HSPs) (Waters 2003). Many HSPs, whose expression is increased by elevated temperatures, assist in the folding of other proteins and prevent irreversible thermal aggregation of proteins (Boston et al. 1996). A review of the different HSPs can be found in Boston et al. (1996).

One interesting group of HSPs are the cytosolic I and II classes of the small HSPs. These are usually expressed in pollen and seed germination, as well as in other developmental stages of the plant (Waters et al. 1996). DeRocher and Vierling (1994) reported the presence of both classes of small HSPs in the cotyledons of the pea seed embryo. They persist in the embryo for four days after imbibition has occurred but disappear shortly after germination. Wehmeyer et al. (1996) reported that cytosolic I small HSPs were most abundant in dry seeds of *Arabidopsis thaliana* but also disappeared shortly after germination (refer also to studies by Larkindale et al. 2005 and Queitsch et al. 2000). The presence of the cytosolic II small HSPs in seeds and their ability to respond to heat stress may indicate a significant role in the evolution of fire-related seed germination responses. Could these cytosolic II small HSPs have coevolved with the same processes in which smoke acts to trigger germination? This is an area of research worthy of further inquiry. The role that small HSPs play may not only have evolved as a result of the fire itself (as suggested by Vogl 1974) but also from the effects of direct sunlight in heating the soil seed bank resulting from clearing the surface vegetation using fire.

There is now evidence to suggest that anthropogenic fires may be currently driving the evolution of seed traits such as seed shape, pubescence, and pericarp thickness. A study was conducted on *Helenium aromaticum*, an annual herb native to the Chilean Mediterranean matorral ecosystem, which, unlike other Mediterranean ecosystems, has not been prone to lightning-ignited fire since the Miocene, but has been increasingly subject to anthropogenic fire since the mid-sixteenth century (Gomez-Gonzalez et al. 2011). From a microevolutionary perspective, Gomez-Gonzalez et al. (2011) found that the seeds of *H. aromaticum* from frequently burned sites were rounder, more densely pubescent, and had thicker pericarps. These traits all enhanced fitness of the species in response to anthropogenic fire. Gomez-Gonzalez et al. (2011) suggest that "the role of fire as a selective agent in ecosystems world-wide may be underestimated."

Mechanism of Action

Given the wide range of species that respond to smoke and its products, it is likely that the active substances act on common metabolic pathways involved in the activation of seed germination. Morphological and physiological processes within the seeds are triggered by environmental cues (e.g., temperature, light, and moisture) during the onset of germination. The correct combination of these cues is essential in order for germination to occur. Dormancy type, morphology of the seeds, and local adaptations to the environment will most likely determine species-specific germination requirements. The network of physiological processes involved in the promotion of germination is still being "untangled" by scientists and will no doubt continue to be challenging.

Many hormones and radicals interact and play an important role in the germination process (Rajjou et al. 2012). These include abscisic acid (ABA), gibberellic acid (GA), ethylene, salicyclic acid, bassinosteroids, auxins, cytokinins, oxylipins, and jasmonic acid. Plant hormone interactions involved in breaking nondeep PD have been reviewed by Kucera et al. (2005) and Finch-Savage and Leubner-Metzger (2006). In seeds of species with nondeep PD, such as those of *Nicotiana tabacum* (Solanaceae) and *Arabidposis thaliana* (Brassicaceae), the dormant embryo is characterized by a high ABA: GA ratio. The endosperm can regulate embryo constraint by influencing this ratio and its sensitivity to these hormones (Bethke et al. 2007; Cadman et al. 2006; Finch-Savage and Leubner-Metzger 2006). Germination of both *A. thaliana* (Pennacchio et al. 2007b) and *N. attenuata* (Preston and Baldwin 1999) occur in response to smoke, and yet until recently neither has previously been used to study the interaction between plant hormones and smoke chemicals (see Nelson et al. 2009).

Attempts also have been made to understand the interaction of smoke and hormones on seeds with nondeep PD (Jäger et al. 1996; Strydom et al. 1996; Thomas and van Staden 1995; van Staden et al. 1995), PD (Kępczyński et al. 2006; Thomas and Davies 2002; van Staden et al. 1995), and PY (Fotheringham et al. 1995; Keeley and Fotheringham 1998b). For example, combinations of N6-benzyladenine (BA; a known germination stimulant) and smoke extract were effective in overcoming induced thermodormancy in lettuce seeds (*Lactuca sativa*; Myricaceae; Strydom et al. 1996). A combination of gibberellin (GA_3) and smoke also effectively broke photodormancy in lettuce seeds (van Staden et al. 1995) and dormant heather seeds (*Calluna vulgaris*; Ericaceae; Thomas and Davies 2002).

Light mediates the biosynthesis of endogenous active GAs, which stimulate α-amylase activity. α- and β-amylases are secreted into the endosperm by the scutellum and aleurone layers of the seeds. One of the roles of these enzymes is to break down the starch contained within the endosperm into oligosaccharides, which are utilized as an energy source by the germinant. Yamaguchi et al. (1998) isolated the gene responsible for this action and proposed a model for gene expression in the GA/light pathway. Incubation of seeds from species that display photodormancy, especially under dark conditions, was shown to inhibit the GA/light pathway. Optimal concentrations of exogenous GAs for inducing germination of *Paulownia tomentosa* (Scrophulariaceae) seeds in the dark were much lower when used with smoke water (Todorović et al. 2005). Smoke water alone did not induce germination of *P. tomentosa* seeds under dark conditions.

Jovanović et al. (2005) synthesized a gibberellic acid nitrite (nGA_3) and tested it on photodormant seeds of *P. tomentosa*. As previously mentioned, nitrogen oxides (NO), which are found in smoke (Keeley and Fotheringham 1997), may interact with members of the phytochrome transduction chain during light-induced germination of *P. tomentosa* (Giba et al. 1998). nGA_3 did not stimulate germination, in contrast to pure GA_3, which did stimulate germination of this species under dark conditions (Jovanović et al. 2005).

Bethke et al. (2007) further teased out the relationship between NO and GA by focusing on the aleurone layer (sole endosperm tissue) in *A. thaliana* seeds. Vacuolation of protein storage vacuoles (PSV) in aleurone cells was significantly less in dormant seeds than in nondormant seeds. Vacuolation was also controlled by temperature. The rate of vacuolation was greatest near the root tip, which is the area that ruptures first during germination. Abscisic acid inhibited vacuolation of PSV's, whereas NO and GA promoted it. The aleurone layer expressed the NO-associated gene (*At*NOS1). The embryo, in contrast, expressed both *At*NOS1 and the GA biosynthetic enzyme, GA_3 oxidase. Nitrogen oxides are important signaling molecules in plants (Parani et al. 2004) and are usually found upstream of GA in the signaling pathway that leads to vacuolation (Bethke et al. 2007). These researchers suggest that aleurone cells are responsible for secreting cell wall degrading enzymes that hydrolyze aleurone cell walls and result in the loss of dormancy. Understanding the role of smoke in the light/GA pathway or germination physiology in general is a pursuit of potentially great promise.

The smoke-derived karrikinolide exhibits similar effects to GA_3 due to its ability to substitute for red light (640 nm) when promoting germination in the Asteraceae of Australia (Merritt et al. 2006). Interestingly, the related karrikins recently discovered in smoke have been shown to trigger germination in *A. thaliana* seeds via a mechanism that requires light and GA_3 synthesis (Nelson et al. 2009). Karrikinolide has also been shown to exhibit cytokinin- and auxin-like activity, especially when applied at low concentrations. A synergistic effect was noted when the application of the substance was combined with kinetin or indole-3-butyric acid (Jain et al. 2008c). A quick glance at the chemical structures of karrikinolide, GA$_3$, and auxins (Figs. 1A, 1H, and 1I, respectively) reveals, however, that significant differences exist in these plant hormones. The germination-promoting properties and chemical structure of karrikinolide more closely resemble those of the strigolactones (Figs. 1B, 1C, 1D, and 1E) (Daws et al. 2008).

The strigolactones are a class of natural sesquiterpene lactones that are potent germination stimulants of the weedy root parasitic plant genera, *Striga*, *Orobanche*, and others (Butler 1995; Joel et al. 1995). These naturally occurring substances appear to be more widely distributed and have greater physiological significance than previously thought (Humphrey and Beale 2006). Examples of strigolactones include orobanchol (Fig. 1E), sorgolactone (Fig. 1F; Sato et al. 2005), strigol (Fig. 1D; Wigchert and Zwanenburg 1999) and the synthetic strigol analogs GR24 (Fig. 1G; GR = Gerry Roseberry from the Johnsons research team) and Nijmegan-1 (Nijmegen, The Netherlands; Daws et al. 2008).

Strigolactone activity seems to reside in the lactone-enol ether D-ring portion, or butenolide moiety (Mangnus 1960), of the molecules (Fig. 1D; Magnus and Zwanenburg 1992; Wigchert and Zwanenburg 1999). These responses in the strigolactone phytohormones are induced by the F-box protein, MAX2, a protein that is also responsible for similar responses of the recently discovered karrikins (Nelson et al. 2011; Waters et al. 2011).

Where karrikinolide is concerned, Flematti et al. (2007) have reported that the methyl substituent in C-3 (Fig. 1A) appears to be important for promoting germination, while substituting C-7 diminishes it. Substituting C-4 and C-5 retains activity. An alkyl substituted analog (3,5-dimethyl-2*H*-furo[2,3-*c*]pyran-2-one) of karrikinolide also is thought to be a contributor to the germination-promoting effects of plant-derived smoke extracts (Flematti et al. 2009). It is common for several related natural products to coexist in plant extracts, some of which work synergistically and others antagonistically. This certainly appears to be the case with smoke, with a number of different germination promoters and inhibitors emerging.

Recently, Jain et al. (2008a) suggested karrikinolide may enhance seed germination and growth by improving water uptake, possibly implicating the involvement of aquaporins. The known aquaporin inhibitors HgCl$_2$ and ZnCl$_2$ reduced seedling water content and affected root development in tomato seeds. The integrity of tomato DNA did not appear to be affected by karrikinolide, but it may play some role during its transcription and translation (Jain et al. 2008b).

These researchers reported that there was up-regulation of genes encoding for expansins. It has been suggested that expansins affect seed germination by weakening the endosperm cap and promoting growth in the embryo of tomato plants (Chen and Bradford 2000). These results offer yet another interesting avenue of inquiry into the promotive effects of smoke and its active substances. Scaffidi et al. (2011) have labeled karrikinolide analogs to evaluate their use in better understanding the germination processes. Clearly, though, more research is still required to determine how smoke-derived compounds mediate their effects.

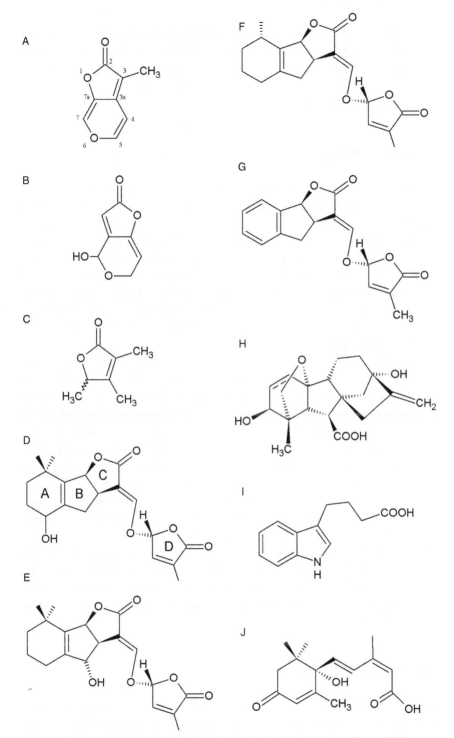

FIGURE 1 Known plant regulators. A. Karrikinolide; B. Patulin; C. 3,4,5-trimethylfuran-2(5H)-one; D. Strigol; E. Orobanchol; F. Sorgolactone; G. GR 24; H. Gibberellic acid (GA$_3$); I. indole-3-butyric acid; J. abscisic acid.

We believe smoke-derived compounds are likely to bind with receptors, with analogs binding to subtypes of those receptors. To date, no relevant receptors have been discovered, but a strigolactone binding protein from membranes of *Striga hemonthica* seeds has (Reizelman et al. 2003) and should perhaps be taken into consideration in future research.

Plant Growth and Vigor

Despite that most of the research has been focused on the effect of smoke on seed germination, it is now becoming clear that smoke is also promoting seedling growth and vigor (Sparg et al. 2005; Chumpookam et al. 2012). Seedling vigor is measured using a vigor index (VI) (Dhindwal et al. 1991), which is calculated as: VI = (shoot length + root length) × percentage germination. The seedling vigor of *Albuca pachychlamys*, *Merwilla natalensis*, and *Tulbaghia violaceae* increased in response to smoke, but interestingly smoke had no significant effect on their germination (Sparg et al. 2005). Smoke water and karrikinolide significantly increased shoot and root elongation of a local rice (*Oryza sativa*) variety (Jain and van Staden 2006; Kulkarni et al. 2006).

Taylor and van Staden (1998) tested the effect of smoke water on the growth of tomato (*Solanum lycopersicum*) roots. Primary root length and secondary root frequency increased significantly in response to specific smoke water dilutions in vitro. The vigor index was significantly greater when karrikinolide-primed tomato seeds were germinated following different seed storage periods, and under different salt concentrations, osmotic potentials, and temperatures (Jain and van Staden 2007). The action of the smoke water did not mimic that of the plant hormones tested (auxins and cytokinins) (Taylor and van Staden 1998). Interestingly, van Staden et al. (2008) reported that harvest indices for tomato and onion (*Allium cepa*) can be improved with smoke water and karrikinolide applications. Chumpookam et al. (2012) discovered that smoke water increased the seedling vigor index and several growth parameters, including root and shoot fresh and dry weights, and length, as well as number of leaves and chlorophyll content of papaya (*Carica papaya*). While it is evident that smoke products promote growth of several crop species, it also clearly demonstrates the potential application of smoke technology for improving crop yields.

Methods for Using Smoke

Seeds and seedlings can be treated with aerosol or liquid smoke solutions. Application of aerosol smoke involves the production of smoke by burning dried plant material (or cellulose products) and pumping smoke into a sealed chamber. The type of plant material used is not important because karrikinolide is produced during the combustion of cellulose and simple carbohydrates within the material, regardless of the quantities or whether or not L-glycine is present (Flemmatti et al. 2011b).

Seeds or punnets of soil containing the seeds are usually placed within the smoke chamber and exposed to smoke for various periods of time. The length of time required is species-specific and usually requires a dose-response like treatment. This is where smoke is applied for progressively greater periods (e.g., 1, 2, 4, 8, 16, 32, and 64 minutes). It is important that the smoke is pumped through a long tube to allow it to cool before it reaches the container. Seed germination can be promoted or inhibited by heat. The chamber should be saturated with a dense smoke. Methods developed by de Lange and Boucher (1990) and Brown et al. (1993) can provide guidance for South African species, Dixon et al. (1995) or Roche et al. (1997a, b) for Australian species, and Keeley and Fotheringham (1998a) for Californian species. All methods are similar, but South African and Western Australian species generally require 30–60 minutes of smoke exposure, while the Californian (United States) species have been shown to require only 5–15 minutes.

Concentrated smoke water is produced by pumping smoke (generated as described above) through deionized water for predetermined periods of time. Smoke water mixtures are prepared in slightly different ways, depending on the amount of water and the length of time the smoke has been pumped through the water. For example, de Lange and Boucher (1990) pumped smoke into 2.5 L of water for 30 minutes (as did Brown 1993b), whereas Taylor and van Staden (1998) vented smoke from 4 kg of leaf material into 500 mL of water for 45 minutes. The concentrated solution is then diluted as required. Seeds are either soaked in the diluted solutions for a certain period of time or the solution is added to the medium (filter paper or soil), in which the seeds are to be tested. The most effective dilution and the exposure time are also species specific. Commercial smoke water products are now available. Karrikinolide is not currently available for purchase, but can be extracted using methods described by Flematti et al. (2004) and van Staden et al. (2004).

Conservation and Land Management and Other Implications

There is great potential for the use of smoke as a practical tool for conservation and land management purposes. Methods for smoke application include priming seeds with smoke prior to sowing (Brown et al. 1998), smoke water application over the soil (Lloyd et al. 2000; Rokich et al. 2002), and aerosol smoke application through the use of tents over small areas of soil (Rokich et al. 2002) or to rooms containing commercial quantities of seeds (Roeder et al. 2011). Brown et al. (1998) showed that priming seeds of the fynbos species *Syncarpha vestita* and *Rhodocoma gigantean* with smoke water followed by drying and storage broke dormancy entirely in *Syncarpha* and partly in *Rhodocoma* and was effective for a period of 12 months (term of the experiment), although its effectiveness in the *Rhodocoma* seeds decreased from 12 weeks onward. Species-specific priming and storage conditions for seeds of a large number of species will need to be developed if this method is to be employed for seed mixtures used in broad-scale restoration.

Smoke water is effective in promoting germination when applied to disturbed areas (Roche et al. 1997 a, b). However, it has not been effective in promoting overall species richness or seedling density (Rokich et al. 2002; Coates 2003). The use of aerosol smoke was more effective than smoke water, but is not practical for broad-scale restoration (Rokich et al. 2002). Norman et al. (2006) found that smoke water was more effective than aerosol smoke when testing a subset of species from the same ecosystem as those tested by Rokich et al. (2002). Seed age, provenance, and environmental factors were attributed as reasons for differences between the two studies. Given that some species are also inhibited by smoke products, this application to every species in the soil seed bank also may not meet the objectives of restoration managers.

Land managers should also take into consideration the effects of smoke and its products on invasive plant species known to inhabit the geographical region of the disturbed or natural areas of interest. The seeds of some invasive exotics are smoke responsive (Read et al. 2000; Adkins and Peters 2001; Daws et al. 2007; Stevens et al. 2007; Franzese and Ghermandi 2011; Djietror et al. 2011; Kandari et al. 2011), while others are not (Smith et al. 1999; Adkins and Peters 2001; Willis et al. 2003; Daws et al. 2007; Stevens et al. 2007). Ruthrof et al. (2011) explored a novel use of smoke water to predict the weed soil seed bank at disturbed sites. All sites had greater seedling density and were dominated by weed species in response to the smoke water treatment compared to the control (Ruthrof et al. 2011). Regular fire regimes or smoke application to the soil seed bank should be used in combination with eradication protocols when the seeds of invasive plant species in the area are smoke responsive. Land managers will need to monitor the responses of both invasive and native plants when developing management protocols for the use of smoke products or fire. Smoke also has the potential to be used in combination with mowing in the maintenance of fragmented grasslands located in urban settings (e.g., Chicagoland areas of Illinois, United States) where fire is no longer available as a tool. The scale of disturbance will determine the most appropriate method to use.

For conservation purposes, much information is now available on the smoke response of plants of rare and endangered species and on those in three of the threatened biodiversity hotspots, all of which are fire-prone Mediterranean ecosystems. For example, Cochrane et al. (2002) have used smoke water solutions extensively to achieve maximum germination of threatened native plant species in Western Australia. Twenty-two percent of the species tested had maximum germination when smoke water was used as a pretreatment. The application of smoke may therefore aid with the successful establishment of rare and endangered species, which require smoke to germinate (Cochrane et al. 2002; Jefferson et al. 2007; Keith 1997; Willis et al. 2003).

Seed of many of these species can be banked, stored and used for restoration purposes with the knowledge that maximum germination will be achieved with the appropriate seed pretreatment. Although scientists are still challenged by the seed germination requirements of many species in ecosystems at risk, our knowledge of the effects of smoke on seed germination has taken us one step further in our quest to conserve the biodiversity of plants in the Cape Floristic Region of South Africa, Southwest Floristic Region of Western Australia, and the Californian Floristic Region of the United States.

The recent isolation and identification of karrikinolide has also opened up another window for broad-scale applications since such small concentrations of this compound are required for germination. Ma et al. (2007) reported that the karrikinolide can, in addition to promoting germination, also stimulate the development and maturation of somatic embryos in *Baloskion tetraphyllum*, a native Australian plant considered important in horticulture. This substance may therefore offer a new phytoactive compound for in vitro culture. This is one of its many applications. It may also find use as a priming agent for triggering repair in aged seeds (Demir et al. 2009), as a weed control agent and for ecological restoration of previously natural sites (Light et al. 2009).

List of Plants

The following is a list of 1,355 species of plants from 120 families whose seeds have been tested for their response to aerosol smoke, smoke water, and plant-derived smoke products, such as karrikinolide (KAR$_1$). Plant species are arranged into monocots and dicots and then alphabetically into families, genera, and species. The binomial names listed are those reported in the original texts unless they have since changed. Where this was the case, the latest accepted name is listed. All nomenclatural authorities and plant names were verified using a number of Internet resources. These include the PlantNet (Botanic Gardens Trust 2013), FloraBase (Western Australian Herbarium 2013), Ecoport (Ecoport 2013), the Plants Database (USDA, NRCS 2013), GRIN (USDA, ARS, National Genetic Resources Program 2013), and IPNI (The International Plant Names Index 2013). All nomenclatural authorities were abbreviated and standardized according to those reported in Brummitt and Powell (1992). The most commonly used vernacular name is also included for each species unless none existed. The conservation status (IUCN 2013) of the listed species has also been included.

Smoke water dilutions have been provided, where appropriate, but should not be compared between studies because of the differences in the materials and methods used in the preparation of these solutions (refer to Methods for Using Smoke above for more detailed information). The methods used in preparing commercial smoke water solutions are not included (e.g., Adkins and Peters 2001; Smith et al. 1999). Readers should be aware that a number of factors determine whether or not a species responds to smoke and should therefore be considered when embarking on studies of this nature. These factors include temperature, seed age, light and dark conditions, scarification, and seed viability, to name a few. The concentration of the smoke and its products are also important, and so it is recommended that different concentrations, doses, and periods of exposure to smoke be tried (see Table 1 in the Introduction for more details).

A total of 18 plant species include line diagrams. All of the illustrations in this book were checked by various botanists and plant experts at the Chicago Botanic Garden and elsewhere. Internet sites, such as GRIN taxonomy and others previously mentioned, were also consulted to check the illustrations. A glossary of definitions is also included to help understand some of the terms used throughout this book.

Monocotyledonae

AGAPANTHACEAE

Agapanthus africanus **Hoffmanns**

AFRICAN BLUE LILY

Seed germination in this species, which occurs in the Cape Floristic Region of South Africa, was not significantly promoted by aerosol smoke (Brown et al. 2003).

ALLIACEAE

Allium cepa **L.**

COMMON ONION

van Staden et al. (2008) reported that common onion seeds treated with smoke water and karrikinolide (KAR_1) produced bulbs with significantly greater diameters and mass and had increased above-ground biomass.

AMARYLLIDACEAE

Cyrtanthus ventricosus **Willd.**

FIRE LILY

This species commonly occurs in the Cape Floristic Region of South Africa. Treatment of its seed with aerosol smoke did not significantly promote germination (Brown et al. 2003).

Dichelostemma pulchellum **(Salisb.) A. Heller**

BLUEDICKS

Germination of bluedicks seeds decreased from 100% (no treatment) to 64% germination when its seeds were treated with charate (0.5 g powdered charred wood per Petri dish) (Keeley and Keeley 1987). Bluedicks are native to the western states of the United States.

ANTHERICACEAE

Agrostocrinum scabrum **(R. Br.) Baill.**

BLUE GRASS LILY

This lily is endemic to the Mediterranean regions of the Western Australian proteaceous heathland, banksia woodlands and jarrah forests. Roche et al. (1997a) reported there was a significant threefold increase in germination of *A. scabrum* seed following smoke water application to the soil seed bank of a rehabilitated bauxite mine site. An aerosol smoke application of 60 minutes duration, prior to sowing, resulted in a significantly greater increase in germination (Roche et al. 1997a). In contrast, Dixon et al. (1995) reported that cold aerosol smoke had no effect on germination while Norman et al. (2006) reported that neither aerosol smoke nor smoke water treatments promoted germination in this species.

Arthropodium milleflorum **(DC.) J. F. Macbr.**

PALE VANILLA LILY

Arthropodium milleflorum is common to the New England region of New South Wales, Australia. Smoke water, when applied on its own or in combination with various other treatments, did not significantly promote germination in this species (Clarke et al. 2000).

Arthropodium strictum **Endl.**

CHOCOLATE LILY

Chocolate lily occurs in open for-ests and grasslands in southeast-ern Australia, namely Victoria, New South Wales, South Australia, and Tasmania. Germination of *A. strictum* doubled from 29% in the control group to 59.3% following ex situ treatments with aerosol smoke (Roche et al. 1997b).

Caesia calliantha **R. Br.**

BLUEGRASS LILY

Seed germination of bluegrass lily seed significantly increased from 4% to 10% when exposed to aerosol smoke for 60 minutes (Roche et al. 1997b). *Caesia calliantha* is native to Australia.

Caesia parviflora **R. Br.**

PALE GRASS LILY

Germination was not significantly improved when seeds of this plant were treated with aerosol smoke (Roche et al. 1997b). This species is widely distributed across Australia.

Arthropodium strictum

Chamaescilla corymbosa (R. Br.) F. Muell. ex Benth.

BLUE SQUILL

Blue squill is widely distributed throughout the Southwest Botanical Province of Western Australia. Smoke water treatments (50% or 100%) of scarified seeds significantly improved germination (Allan et al. 2004). In contrast, smoke water had no effect on it when applied to the surface soil of a rehabilitated bauxite mine in the southwest of Western Australia (Roche et al. 1997a). Aerosol smoke seemed equally ineffective on broadcast seeds from mine soils (Roche et al. 1997a). In other studies, aerosol smoke treatment of the soil surface did, however, significantly increased germination in this species (Dixon et al. 1995; Roche et al. 1997b).

Johnsonia lupulina R. Br.

HOODED LILY

Cold aerosol smoke treatments had no effect on the germination of this Western Australian plant (Dixon et al. 1995).

Laxmannia omnifertilis Lindl.

Laxmannia omnifertilis commonly occurs in mixed eucalypt and banksia woodlands and jarrah forests in the Southwest Botanical Province of Western Australia. A significant increase in final germination percentage was reported when this species was treated with aerosol smoke for 60 minutes (Roche et al. 1997b). Smoke water sprays, with concentrations of 50 and 100 mL/m^2, in contrast, had no effect on the germination of seeds scattered in an intact banksia woodland 20 km south of Perth, Western Australia (Lloyd et al. 2000).

Laxmannia orientalis Keighery

DWARF WIRE-LILY

Germination of dwarf wire-lily seeds, collected from a *Eucalyptus baxteri* heathy-woodland in Victoria, Australia, was significantly promoted following a combined heat and smoke water treatment (Enright and Kintrup 2001).

Sowerbaea laxiflora Lindl.

PURPLE TASSELS

Germination of this native Western Australian shrub, which occurs in proteaceous heathland, banksia woodland, and jarrah forest communities, increased significantly when seeds were treated with high temperature (40°C for 90 days), sown in soil, and then exposed to aerosol smoke for 60 minutes (Tieu et al. 2001a).

Sowerbaea multicaulis E. Pritz. in Diels.

MANY-STEMMED LILY

Cold smoke treatment of the seeds in a smoke tent for 90 minutes had no effect on the germination of this Western Australian plant (Dixon et al. 1995).

Thysanotus dichotomus **(Labill.) R. Br.**

Germination of the seeds of this Australian species, which occurs in banksia woodlands of Western Australia, was not significantly promoted when they were soaked for 24 hours in 10% smoke water (Clarke and French 2005).

Thysanotus fastigiatus **Brittan.**

Thysanotus fastigiatus is native to the banksia woodland and jarrah forest communities of Western Australia. No significant response to the germination of the seeds of this species was observed when smoke water and aerosol smoke were applied in situ to the soil surface of jarrah forest communities (Roche et al. 1997a). Norman et al. (2006) reported similar findings.

Thysanotus multiflorus **R. Br.**

MANY FLOWERED FRINGE LILY

Native to the proteaceous heathlands, banksia woodlands, and jarrah forests of the Southwest Botanical Province of Western Australia, the germination of this lily doubled in response to smoke water treatment of the soil seed bank (Roche et al. 1997a). In addition, the germination of this species significantly increased following both in situ (Roche et al. 1997a) and ex situ (Dixon et al. 1995; Roche et al. 1997b) aerosol smoke applications. Norman et al. (2006) reported, in contrast, that neither aerosol smoke nor smoke water treatments significantly promoted germination in this species.

Thysanotus thyrsoideus **Baker**

COMMON FRINGE LILY

A 1 L/m² application of undiluted smoke water applied to the soil surface of a rehabilitated bauxite mine in the southwestern region of Western Australia had no significant effect on the germination of the seeds of this species (Roche et al. 1997a).

Tricoryne elatior **R. Br.**

YELLOW RUSH LILY

Yellow rush lily occurs throughout Australia. Aerosol smoke treatments of broadcast seeds in rehabilitated mine sites in the southwestern region of Western Australia had no effect on their germination (Roche et al. 1997a,b).

Tricoryne humilis **Endl.**

This species occurs in the jarrah forests of Western Australia. Neither aerosol smoke nor smoke water treatments significantly promoted germination of its seeds (Norman et al. 2006).

ARALIACEAE

Aralia cordata **Thunb.**

JAPANESE SPIKENARD

The effects of aerosol smoke, heat, darkness, cold stratification, and combinations of smoke with each of the three other treatments on seed germination were examined in this study (Tsuyuzaki and Miyoshi 2009). Smoke was produced by burning Timothy hay (*Phleum pratense*), which was pumped through a 3.5 m cooling tube into a smoke chamber for approximately 5 minutes. The seeds were exposed to the smoke for 60 minutes. Those seeds exposed also to heat were incubated at 75°C for 25 minutes. The cold stratification process took 1 month, during which the seeds remained in an incubator set at 4°C. Where the dark treatment was concerned, the seeds were maintained in total darkness for the entire germination period. The smoke and dark treatments, as well as the combinations of smoke with heat, smoke and dark, and smoke with cold stratification all significantly inhibited germination. This species is native to Japan, Korea, and China.

Aralia elata **(Miq.) Seem.**

JAPANESE ANGELICA TREE

The effects of aerosol smoke, heat, darkness, cold stratification, and combinations of smoke with each of the three other treatments on seed germination were examined in this study (Tsuyuzaki and Miyoshi 2009; see *A. cordata* above for details about the tests performed). Germination was significantly inhibited when the seeds were treated with aerosol smoke alone. When the smoke treatment was combined with heat (75°C for 25 minutes), it was significantly increased. This species did not germinate at all when incubated in darkness.

ASPHODELACEAE

Asphodelus ramosus **L.**

COMMON ASPHODEL

Crosti et al. (2006) reported that exposure to cooled aerosol smoke for 60 minutes significantly inhibited germination of the seeds of this species. *Asphodelus ramosus* is common to parts of Europe and Africa.

Bulbine bulbosa **(R. Br.) Haw.**

BULBINE LILY

Bulbine lily is common to the New England region of New South Wales, Australia. Smoke water, when applied on its own or in combination with other treatments to the seeds of this species, did not significantly promote germination (Clarke et al. 2000).

Kniphofia uvaria **(L.) Oken**

RED HOT POKER

RED HOT POKER

Germination of red hot poker seed, a species which occurs in the Cape Floristic Region of South Africa, was not significantly promoted by aerosol smoke (Brown et al. 2003).

Trachyandra **sp.**

Germination of an unidentified species of *Trachyandra*, commonly found in the Cape Floristic Region of South Africa, was not significantly promoted by aerosol smoke (Brown et al. 2003).

CAESALPINIACEAE

Cassia mimosoides **L.**

FEATHER-LEAVED CASSIA

Seeds collected from a Sudanian savanna-woodland in Burkina Faso, Africa, were treated with a variety of fire cues to determine their effects on seed germination (Dayamba et al. 2010). The seeds were soaked in smoke water (at concentrations of 100%, 75%, 50%, 25%, and 5% of the stock solution) for 24 hours. The smoke water stock solution was produced by burning a mixture of dominant native species from the Tiogo and Laba State forests of Sudan, and pumping the smoke through water for 10 hours. The seeds also underwent a heat shock treatment, during which they were incubated in an oven at 40, 80, 120, or 140°C for 2.5 minutes. Following these treatments, germination percentages and mean germination times were measured. None of the smoke or heat shock treatments had any effect on the germination of this species.

CALYCERACEAE

Boopis gracilis **Phil.**

BOOPIS

The germination of *B. gracilis* seeds, when treated with a combination of heat (80°C for 5 minutes) and aerosol smoke for 60 minutes increased, but was not significant ($P<0.05$) (Gonzalez and Ghermandi 2012). This species is native to the grasslands of northwestern Patagonian.

CARICACEAE

Carica papaya **L. cv. Tainung No. 2**

PAPAYA

Germination, growth and anatomical structure of papaya seeds were studied in response to smoke water treatments (Chumpookam et al. 2012). Smoke water was prepared by burning dry rice straw (*Oryza sativa*) and bubbling the smoke through 500 mL of distilled water for 45 minutes. Papaya seeds were soaked with different concentrations of smoke water (0.1%, 0.2%, 1%, 2%, 3%, 4%, 5%, 7%, or 10% [v/v]) for 24 hours, rinsed and were then sowed. Germination rate and average germination times were measured. Smoke water treatments of 0.1% and 0.2% v/v significantly increased germination rates (87.3% and 86.7%, respectively). No other concentrations of smoke water had any effect on germination rates or significantly altered average germination times. In the growth experiments, papaya seedlings at the two true leaf stage were transplanted (four plants/pot) into peatmoss-filled pots, which were saturated with one of eight different concentrations of smoke water (0%, 1%, 2%, 3%, 4%, 5%, 7%, or 10% [v/v]). These were then incubated in a growth chamber and irrigated three times per week with water for a period of 4weeks. After 4 weeks, growth parameters (shoot length (cm), root length (cm) shoot fresh weight (g), shoot dry weight (g), root fresh weight (g), root dry weight (g), number of leaves, chlorophyll content and seedling vigor index) and mineral content (nitrogen, phosphorus, potassium, calcium, magnesium, iron, manganese, zinc, and copper) of roots and shoots were measured. All growth parameters measured increased significantly in response to the smoke water treatments. Where mineral content was concerned, nitrogen was the only mineral in the roots that increased significantly in response to smoke treatments. All other minerals in the roots remained the same. Nitrogen and magnesium content in the shoots increased significantly when the seedlings were treated with 5%, 7%, or 10% v/v smoke water solutions. None of the other minerals responded to the smoke treatments. Papaya seeds were then soaked for 24 hours in one of six concentrations of smoke water (0%, 0.1%, 0.2%, 1%, 3%, or 10% [v/v]), after which they were rinsed with distilled water prior to studying their anatomical structure. Seeds were embedded in a tissue-freezing medium, cut into 30-mm sections, stained using 0.1% (w/v) of safranin and then observed under a microscope. Longitudinal sections of papaya seeds treated with smoke water revealed that low concentrations of smoke water (0.1% and 0.2%) stimulated the endosperm to stretch and the seed coat to rupture thus allowing the radicle to elongate and emerge.

CENTROLEPIDACEAE

Centrolepis aristata **(R. Br.) Roem. & Schult.**

POINTED CENTROLEPIS

Germination of *C. aristata* seeds collected from a *Eucalyptus baxteri* heathy-woodland in Victoria, Australia, significantly improved following a combined heat and smoke water treatment (Enright and Kintrup 2001).

Centrolepis strigosa **(R. Br.) Roem. & Schult.**

HAIRY CENTROLEPIS

Like *C. aristata* above, germination of hairy centrolepis seed, collected from a heathy-woodland in Victoria, Australia, was significantly promoted by a combination of heat and smoke water treatments (Enright and Kintrup 2001). In another study, soil samples from the Eden Burning Study Area, a dry sclerophyll forest in the Yalumba State Forest of New South Wales, Australia, were collected and air dried to test the effects of heat, smoke, and an interaction between the two cues on seeds from the seed bank (Penman et al. 2008). Samples exposed to heat treatment were incubated at 80°C for 60 minutes while those exposed to smoke were incubated in a room, where smoke was generated for 120 minutes. Only the heat treatment significantly improved germination of the seeds of this species. Neither the smoke treatment nor the interaction between the two cues had any effect.

COCHLOSPERMACEAE

Cochlospermum planchonii **Hook. f. ex Planch.**

FALSE COTTON

Seeds collected from a Sudanian savanna-woodland in Burkina Faso, Africa, were treated with a variety of fire cues to determine their effects on seed germination (Dayamba et al. 2010). The seeds were soaked in smoke water (at concentrations of 100%, 75%, 50%, 25%, and 5% of the stock solution) for 24 hours. The smoke water stock solution was produced by burning a mixture of dominant native species from the Tiogo and Laba State forests of Sudan, and pumping the smoke through water for 10 hours. The seeds also underwent a heat shock treatment, during which they were incubated in an oven at 40, 80, 120, or 140°C for 2.5 minutes. Following these treatments, germination percentages and mean germination times were measured. None of the smoke or heat shock treatments had any effect on the germination of this species.

COLCHICACEAE

Burchardia umbellata **R. Br.**

MILKMAIDS

Milkmaids are native to the southwestern region of Western Australia, specifically the proteaceous heathlands, banksia woodlands, and jarrah forests. Seed germination increased significantly in response to a 90-minute ex situ treatment of aerosol smoke (Dixon et al. 1995). Germination was also enhanced in response to aerosol smoke treatment for a 60 minute period prior to and after sowing (Roche et al. 1997a,b). Bell et al. (1987) similarly reported that treatment with charate promoted germination in this species. Smoke water had no effect, however (Roche et al. 1997a). Interestingly, Norman et al. (2006) reported that neither aerosol smoke nor smoke water treatments had any effect on the germination of this species.

COMMELINACEAE

Aneilema acuminatum **R. Br.**

POINTED ANEILEMA

The pointed aneilema is native to Australia, occurring mainly in the states of Queensland and New South Wales. Aerosol smoke treatments of 60 minutes duration on soil samples containing seeds of this species, collected across forest edges between subtropical rainforests and eucalypt forests in the Lamington National Park of Queensland, Australia, did not promote germination (Tang et al. 2003).

Pollia crispata **(R. Br.) Benth.**

POLLIA

Like *Aneilema acuminatum* above, *Pollia crispata* is native to Australia, and also occurs in Queensland and New South Wales. Soil samples collected in the Lamington National Park of Queensland, Australia, were exposed to cool aerosol smoke for 60 minutes without any improvement to seed germination (Tang et al. 2003).

Tradescantia ohiensis **Raf.**

BLUE JACKET

Blue jacket seed, collected from a tallgrass prairie in the Midwest of the United States, did not germinate in response to aerosol smoke treatments of 1, 10, or 60 minutes duration (Jefferson et al. 2007).

CUNONIACEAE

Callicoma serratifolia **Andrews**

BLACK WATTLE

Callicoma serratifolia is a monotypic genus that is native to Australia. Sixty minutes of aerosol smoke treatments of soil samples containing the seeds of this species did not significantly promote germination (Tang et al. 2003).

CYPERACEAE

Baumea nuda **(Steud.) S. T. Blake**

This species occurs throughout the east coast of Australia. A 60-minute aerosol smoke treatment of soil samples containing the seeds of this sedge from subtropical rainforest and eucalypt forests in Queensland, Australia, did not improve their germination (Tang et al. 2003).

Carex inversa **R. Br.**

KNOB SEDGE

Knob sedge is common to the New England region of New South Wales, Australia. Smoke water, when applied on its own or in combination with other treatments, did not significantly promote germination in this species (Clarke et al. 2000).

Carex oxyandra **Kudô**

SEDGE

The effects of aerosol smoke, heat, darkness, cold stratification, and combinations of smoke with each of the three other treatments on seed germination were examined in this study (Tsuyuzaki and Miyoshi 2009). Smoke was produced by burning Timothy hay (*Phleum pratense*), which was pumped through a 3.5 m cooling tube into a smoke chamber for approximately 5 minutes. The seeds were exposed to the smoke for 60 minutes. Those seeds exposed also to heat were incubated at 75°C for 25 minutes. The cold stratification process took 1 month, during which the seeds remained in an incubator set at 4°C. Where the dark treatment was concerned, the seeds were maintained in total darkness for the entire germination period. None of the treatments had any effect on the germination of this species.

Carex **sp.**

A 60-minute aerosol smoke treatment of soil samples containing seeds of an unidentified *Carex* species, collected across forest edges between subtropical rainforests and eucalypt forests in the Lamington National Park of Queensland, Australia, did not significantly promote germination (Tang et al. 2003).

Caustis flexuosa **R. Br.**

CURLY WIG

Soil samples from the Eden Burning Study Area, a dry sclerophyll forest in the Yalumba State Forest of New South Wales, Australia, were collected and air dried to test the effects of heat, smoke, and an interaction between the two cues on seeds from the seed bank. Samples exposed to heat treatment were incubated at 80°C for 60 minutes while those exposed to smoke were incubated in a room, where smoke was generated for 120 minutes. Only the smoke treatment significantly improved the germination of the seeds of this species (Penman et al. 2008).

Cyathochaeta avenacea **(R. Br.) Benth.**

Cyathochaeta avenacea occurs in proteaceous heathland, banksia woodland, and jarrah forest regions of Western Australia. Germination significantly increased in response to aerosol smoke treatment of the soil seed bank on the disturbed soils of a bauxite mine site, but smoke water had no effect (Roche et al. 1997a). Norman et al. (2006) reported that neither aerosol smoke nor smoke water treatments promoted germination in this species.

Cyperus aquatilis **R. Br.**

FLAT SEDGE

Flat sedge occurs throughout Australia. A 60-minute aerosol smoke treatment of soil samples containing seeds of this species, collected in both subtropical rainforest and eucalypt forests in the Lamington National Park of Queensland, Australia, resulted in no germinants (Tang et al. 2003).

Cyperus enervis **R. Br.**

CYPERUS SEDGE

Like *C. aquatilis* above, the Cyperus sedge also occurs on the east coast of Australia. A 60-minute aerosol smoke treatment of soil samples from the Lamington National Park of Queensland, Australia, did not stimulate germination of its seeds (Tang et al. 2003).

Cyperus gracilis **R. Br.**

SLENDER SEDGE

Aerosol smoke treatments of 60 minutes duration applied to soil seed banks in an open-cut coal mine of the Hunter Valley in New South Wales, Australia, did not significantly improve or promote germination of its seeds (Read et al. 2000). This species commonly occurs in the eastern states of Australia.

Cyperus tetraphyllus **R. Br.**

A 60 minute aerosol smoke treatment of soil samples collected in subtropical rainforest and eucalypt forests of Queensland, Australia, did not significantly affect germination (Tang et al. 2003). The distribution of this sedge is confined to the east coast of Australia.

Eleocharis keigheryi **K. L. Wilson**

Maximum germination of this threatened and endangered species, which is native to the southwestern proteaceous heathland, banksia woodland, and jarrah forest communities of Western Australia, was achieved by removal of the seed coat, followed by adding gibberellic acid to the growth medium (25 mg/L) and by soaking the seeds in smoke water for 24 hours (Cochrane et al. 2002).

Gahnia clarkei **Benth.**

TALL SAW-SEDGE

The tall saw-sedge is native to Australia and occurs in several states. Freshly collected seeds were treated for 60 minutes with cool aerosol smoke, some of which had received other treatments, but germination was not significantly promoted (Roche et al. 1997b). In another study, soil samples from the Eden Burning Study Area, a dry sclerophyll forest in the Yalumba State Forest of New South Wales, Australia, were collected and air dried to test the effects of heat, smoke, and an interaction between the two cues on seeds from the seed bank (Penman et al. 2008). Samples exposed to heat treatment were incubated at 80°C for 60 minutes while those exposed to smoke were incubated in a room, where smoke was generated for 120 minutes. None of the treatments had any effect on the germination of the seeds of this species.

Gahnia decomposita **(R. Br.) Benth.**

LITTLE SEDGE

Dixon et al. (1995) reported that 90 minutes of cold smoke treatment had no effect on seed germination for this Western Australian sedge species.

Gahnia lanigera **(R. Br.) Benth.**

BLACK GRASS SAW SEDGE

Gahnia lanigera occurs commonly in coastal regions of southwest, Western Australia. Germination of this sedge's seeds increased from 0.6% to 14.5% when treated with aerosol smoke (Roche et al. 1997b).

Gahnia radula **(R. Br.) Benth.**

THATCH SAWSEDGE

Soil samples from the Eden Burning Study Area, a dry sclerophyll forest in the Yalumba State Forest of New South Wales, Australia, were collected and air dried to test the effects of heat, smoke, and an interaction between the two cues on seeds from the seed bank. Samples exposed to heat treatment were incubated at 80°C for 60 minutes while those exposed to smoke were incubated in a room, where smoke was generated for 120 minutes. None of the treatments had any effect on the germination of the seeds of this species (Penman et al. 2008).

Gahnia sieberiana **Kunth**

RED FRUITED SAW SEDGE

Red fruited saw sedge grows in wet areas and has a broad distribution along the eastern coastline of Australia. It also occurs in New Guinea and New Caledonia. In a study into the effects of heat shock and smoke on seed germination, Thomas et al. (2003) reported that this species responded best, with a maximum germination of 53%, to a heat shock treatment of 100°C for 5 minutes. An aerosol smoke treatment of 20 minutes duration achieved 38% germination, in comparison to 26% for the control.

Isolepis marginata **(Thunb.) A. Dietr.**

LITTLE CLUB-RUSH

This plant is native to South Africa and is invasive in Australia and New Zealand. Germination of its seeds was promoted by a smoke water treatment of the soil seed bank (Enright and Kintrup 2001).

Lepidosperma angustatum **R. Br.**

PITHY SWORD-SEDGE

Dixon et al. (1995) reported that 90 minutes of cold aerosol smoke treatment had no effect on seed germination of this Western Australian species. A 60-minute treatment of cold aerosol smoke, when applied alone and in combination with other treatments, also had no effect (Roche et al. 1997b).

Lepidosperma gladiatum **Labill.**

COAST SWORD-SEDGE

Like *L. angustatum*, Dixon et al. (1995) reported that 90 minutes of cold smoke treatment had no effect on seed germination of this Western Australian species.

Lepidosperma laterale **R. Br.**

VARIABLE SWORD SEDGE

Soil samples from the Eden Burning Study Area, a dry sclerophyll forest in the Yalumba State Forest of New South Wales, Australia, were collected and air dried to test the effects of heat, smoke, and an interaction between the two cues on seeds from the seed bank. Samples exposed to heat treatment were incubated at 80°C for 60 minutes while those exposed to smoke were incubated in a room, where smoke was generated for 120 minutes. The heat treatment significantly increased germination, while the increase due to aerosol smoke was only marginally significant (Penman et al. 2008). The interaction between the two cues had no effect.

Lepidosperma longitudinale **Labill.**

PITHY SAW-SEDGE

This sedge is native to Australia and occurs throughout the country. Freshly collected seeds were treated with cool aerosol smoke for 60 minutes, but germination was not significantly promoted (Roche et al. 1997b).

Loxocarya striatus **F. Muell.**

Loxocarya striatus is native to the proteaceous heathlands, banksia woodlands, and jarrah forests of southwestern Western Australia. Tieu et al. (2001a) revealed that smoke promoted germination, but high temperature treatments of the seeds were more effective.

Mariscus thunbergii **(Vahl.) Schrad.**

Mariscus thunbergii occurs in the Cape Floristic Region of South Africa. Aerosol smoke did not significantly promote germination of the seeds of this species (Brown et al. 2003).

Sceleria tricuspidata **S. T. Blake**

This species occurs naturally on the east coast and Northern Territory of Australia. Sixty minutes of aerosol smoke treatment of soil samples from Queensland, Australia, did not promote or inhibit germination in this species (Tang et al. 2003).

Schoenus apogon **Roem. & Schilt.**

COMMON BOG RUSH

Soil samples from the Eden Burning Study Area, a dry sclerophyll forest in the Yalumba State Forest of New South Wales, Australia, were collected and air dried to test the effects of heat, smoke, and an interaction between the two cues on seeds from the seed bank. Samples exposed to heat treatment were incubated at 80°C for 60 minutes while those exposed to smoke were incubated in a room, where smoke was generated for 120 minutes. Only the heat treatment significantly improved the germination of the seeds of this species (Penman et al. 2008). Neither smoke nor the interaction between the two cues had any effect.

Schoenus maschalinus **Roem. & Schilt.**

LEAFY BOG RUSH

Like *S. apogon* above, soil samples from the Eden Burning Study Area, a dry sclerophyll forest in the Yalumba State Forest of New South Wales, Australia, were collected and air dried to test the effects of heat, smoke, and an interaction between the two cues on seeds from the seed bank. Samples exposed to heat treatment were incubated at 80°C for 60 minutes while those exposed to smoke were incubated in a room, where smoke was generated for 120 minutes. None of the treatments had any effect on the seeds of this species (Penman et al. 2008).

Schoenus melanostachys **R. Br.**

BLACKHEAD SEDGE

This species occurs naturally on the east coast and southeast of Australia. A 60 minute aerosol smoke treatment of soil samples from the Lamington National Park of Queensland, Australia, did not significantly promote germination (Tang et al. 2003).

Tetraria capillaris **(F. Muell.) J. M. Black**

HAIR SEDGE

Tetraria capillaris occurs naturally along the southern coast line of Australia. A 1 L/m² application of undiluted smoke water to the soil surface of a rehabilitated bauxite mine in the southwest of Western Australia had no significant effect on the germination of seeds collected from the soil seed bank (Roche et al. 1997a).

Tetraria octandra **(Nees) Kük.**

This species occurs in the southwestern region of Western Australia. Freshly collected seeds were treated with cool aerosol smoke for 60 minutes, some of which had also received other treatments, but germination was not significantly promoted (Roche et al. 1997b).

DASYPOGONACEAE

Dasypogon bromeliifolius **R. Br.**

DRUMSTICKS

This species occurs in the Perth and southwestern region of Western Australia. Freshly collected seeds were treated with cool aerosol smoke for 60 minutes without any significant effects on germination (Roche et al. 1997b).

Kingia australis **R. Br.**

BULLANOCK

This monotypic species of the genus, *Kingia*, occurs only in the southern half of Western Australia and bears a close resemblance to grasstrees (*Xanthorrhoea* spp.). Freshly collected seeds were treated with 60 minutes of cool aerosol smoke, but germination was not significantly promoted (Roche et al. 1997b). Bullanock is its Australian Aboriginal name, which is considered more appropriate than its previous vernacular, "Black gin."

LINACEAE

Linum suffruticosum **L.**

WHITE FLAX

The seeds of this species were incubated for 24 hours in two smoke water solutions of 1:1 and 1:10 concentrations, prepared according to Jager et al. (1996b). The smoke water solutions had no effect on germination percentage or rate, but improved seedling growth (Moreira et al. 2010). This species is common throughout the Mediterranean Basin and is native to the southern parts of Europe and northern Africa.

LOMANDRACEAE

Lomandra integra **T. D. MacFarl.**

Freshly collected seeds of this Australian native did not significantly respond to cool aerosol smoke treatments of 60 minutes duration (Roche et al. 1997b).

Lomandra longifolia **Labill.**

SPINY-HEADED MAT RUSH

Spiny-headed mat rush is common to the New England region of New South Wales, Australia. Smoke water treatments did not significantly promote germination in this species (Clarke et al. 2000).

Lomandra micrantha **(Endl.) Ewart**

SMALL-FLOWERED MAT-RUSH

Like *L. integra*, the freshly collected seeds of this Australian native were treated with 60 minutes of cool aerosol smoke, but germination was not significantly promoted (Roche et al. 1997b).

Lomandra multiflora **(R. Br.) Britten**

MANY-FLOWERED MAT-RUSH

This rush is native to Western Australian proteaceous heathlands, banksia woodlands, and jarrah forests. In a study into the application of smoke water to the soil seedbank of mine sites, Roche et al. (1997a) reported that *L. multiflora* was one of 20 species whose germination increased significantly. Ex situ studies also revealed that germination increased when seeds were treated with aerosol smoke after they had been sown (Roche et al. 1997b). Germination was not, however, significantly affected by smoke water treatment of the soil seed bank of bauxite mines (Roche et al. 1997a).

Lomandra nigricans **T. D. MacFarl.**

SILKY MAT-RUSH

Like *L. integra* and *L. micrantha*, the freshly collected seeds of this Australian native were treated for 60 minutes with cool aerosol smoke, some of which had received other treatments, but germination was not significantly promoted (Roche et al. 1997b).

Lomandra preissii **(Endl.) Ewart**

This species occurs in the Southwest Floristic Region of Western Australia. Merritt et al. (2007) used warm stratification (26/13°C or 33/18°C) and smoke water to break dormancy and promote germination of *L. preissii*. Norman et al. (2006) reported, in contrast, that neither aerosol smoke nor smoke water treatments significantly promoted germination of its seed.

Lomandra sericea **(Endl.) Ewart**

SILKY MAT RUSH

A 1 L/m^2 application of undiluted smoke water to the soil surface of a rehabilitated bauxite mine in the southwest of Western Australia had no significant effect on the germination of seeds from the soil seed bank (Roche et al. 1997a).

Lomandra sonderi **(F. Muell.) Ewart**

This species occurs in the jarrah (*Eucalyptus marginata*) forests of Western Australia. Neither aerosol smoke nor smoke water treatments (Roche et al. 1997a) significantly promoted germination of its seeds (Dixon et al. 1995; Roche et al. 1997b; Norman et al. 2006).

Lomandra **sp.**

Treatment with smoke water significantly promoted germination in the seeds of an unidentified species of *Lomandra*, which occurs in the jarrah (*Eucalyptus marginata*) forests of Western Australia (Roche et al. 1997a). Aerosol smoke treatments of broadcast seeds had no effect.

Lomandra spartea **(Endl.) Ewart**

Like other *Lomandra* species, aerosol smoke had no effect on the germination of this Australian native plant (Roche et al. 1997b).

DIOSCOREACEAE

Dioscorea dregeana **(Kunth) T. Durand & Schinz**

WILD YAM

Both smoke water (1:500 v/v) and smoke-derived karrikinolide (10^{-7} M), significantly promoted germination in this tuberous medicinal plant species and also enhanced seedling vigor (Kulkarni et al. 2007b).

ECDEIOCOLEACEAE

Ecdeiocolea monostachya **F. Muell.**

This grasslike species occurs in Western Australia. Freshly collected seeds were treated for 60 minutes with cool aerosol smoke, some of which had received other treatments, but germination was not significantly promoted (Roche et al. 1997b).

Georgiella hexandra **B. G. Briggs & L. A. Johnson.**

This species occurs in the Southwest Botanical Province of Western Australia. A significant increase in final germination percentage was observed when *G. hexandra* seed had been treated with aerosol smoke for 60 minutes (Roche et al. 1997b). The seeds received the treatment shortly after being sown. The best results were achieved when seeds were stored in soil for 12 months and then treated with smoke. Final germination in those studies was 21.1% (Roche et al. 1997b).

HAEMODORACEAE

Anigozanthos bicolor **Endl.**

LITTLE KANGAROO PAW

This species is a native of southwestern Western Australia's proteaceous heathland, banksia woodlands, and jarrah forest regions. Germination was significantly improved when the soil seed bank was exposed to aerosol smoke for 90 minutes (Dixon et al. 1995).

Anigozanthos flavidus **DC.**

EVERGREEN KANGAROO PAW

The effects of "Seed Starter" smoke water, karrikinolide (KAR₁), glyceronitrile, and cellulose-derived smoke water were tested on the seeds of this species (Downes et al. 2013). Germination was promoted by "Seed starter" smoke water and to some degree by the cellulose-derived smoke water. Concentrations of glyceronitrile, ranging from 1 to 500 µM also stimulated germination, with the greatest increases occurring at concentrations from 25 to 300 µM. Seeds treated with concentrations of 400 µM or greater inhibited germination. KAR₁ had no effect, suggesting that other chemicals in smoke promote germination. This species occurs in the southwest of Western Australia.

Anigozanthos humilus **ssp.** chrysanthus **Hopper**

GOLDEN CATSPAW

Like *A. bicolor*, this species is native to Western Australia's proteaceous heathland, banksia woodlands, and jarrah forest regions. Germination was significantly promoted when the soil seed bank was treated with aerosol smoke for 90 minutes (Dixon et al. 1995). This species is considered vulnerable by the IUCN (2013).

Anigozanthos manglesii **D. Don**

MANGLE'S KANGAROO PAW

This native of the Southwest Botanical Province of Western Australia, and floral emblem of the state, germinated more readily when exposed to aerosol smoke (Dixon et al. 1995; Roche et al. 1997b). Tieu et al. (2001a) reported that heat shock (30 minutes, 3 hours or 24 hours) at 100°C was, however, more effective in promoting seed germination in comparison to a smoke treatment of 60 minutes. Treating seeds with smoke, followed by heat shock treatment, provided the best germination results for this species. Tieu et al. (2001b) also reported that seeds required 360 days of shelf storage and then smoke treatment in order to achieve 30% germination. Seed storage alone did not improve

Anigozanthos manglesii

germination. Tan (2005) reported that smoke water (10% for 24 hours) was ineffective in promoting germination of this species, regardless of seed age (0.5, 1.5, and 3.5 years since collection). Heat (80°C for 120 minutes) resulted in increased germination (90%). Norman et al. (2006) reported that neither aerosol smoke nor smoke water treatments significantly promoted germination of its seeds.

Anigozanthos rufus **Labill.**

RED KANGAROO PAW

Anigozanthos rufus is common in the jarrah forests, mallee woodlands and proteaceous heathlands of Western Australia's southwest. Germination of its seeds was significantly promoted by aerosol smoke (Roche et al. 1997b).

Blancoa canescens **Lindl.**

WINTER BELL

Winter bell is native to the banksia woodlands and proteaceous scrub-heaths of Western Australia. High germination percentages occurred in response to smoke treatments and 47.7% final germination when sown seeds had been treated with smoke for 60 minutes (Roche et al. 1997b).

Conostylis aculeata **R. Br.**

PRICKLY CONOSTYLIS

Conostylis aculeata is native to Western Australia's proteaceous heathland, banksia woodland, and jarrah forest communities. Norman et al. (2006) reported that in situ germination of this species increased in response to aerosol smoke treatment, but Roche et al. (1997a) reported there was no effect on the germination of broadcast seeds in rehabilitated mine soils. Karrikinolide significantly increased germination at concentrations ranging from 1 ppm to 1 ppt (Flematti et al. 2004). Germination was equivalent or higher than that obtained by treating the seed with smoke water (1% or 10%).

Conostylis candicans **Endl.**

GREY COTTONHEAD

This species is native to Western Australia's proteaceous heathland, banksia woodland, and jarrah forest communities. Lloyd et al. (2000) reported that ex situ smoke water (1 L/m^2) and 60 minutes of aerosol smoke treatment significantly improved seed germination. Aerosol smoke treatment achieved the highest germination (20%) of all treatments. Flematti et al. (2004) showed that karrikinolide (10 ppb) also promoted germination in this species.

Conostylis neocymosa **Hopper**

Conostylis neocymosa is a native of southwestern Western Australia's proteaceous heathland, banksia woodland, and jarrah forest regions. Ex situ germination improved when the soil seed bank was exposed to aerosol smoke for 90 minutes

(Dixon et al. 1995). Tieu et al. (2001b) showed that a combination of seed burial in soil for 450 days and then smoke treatment for 60 minutes achieved 75% germination in dormant seeds.

Conostylis serrulata **R. Br.**

Germination of *C. serrulata*, also a native of southwestern Western Australian proteaceous heathland, banksia woodlands, and jarrah forest communities, increased in response to smoke water treatment of the soil seed bank on the disturbed soils of a bauxite mine site (Roche et al. 1997a).

Conostylis setigera **R. Br.**

BRISTLY COTTONHEAD

Bristly cottonhead is common throughout the Southwest Botanical Province of Western Australia. Tieu et al. (1999) reported that germination of this species significantly increased in response to imbibition in smoke water.

Conostylis setosa **Lindl.**

WHITE COTTONHEAD

White cottonhead occurs in the southwestern Western Australian proteaceous heathland, banksia woodland, and jarrah forest communities. This is a smoke-responsive species that has exhibited positive germination results when treated with aerosol smoke, both in situ (Roche et al. 1997a) and ex situ (Dixon et al. 1995). Roche et al. (1997a) reported, in contrast, that aerosol smoke treatments of broadcast seeds in rehabilitated mine sites in the southwest of Western Australia had no significant effect on germination.

Dilatris pillansii **W. F. Barker**

ROOIWORTEL (AFR.)

Seed germination in this fynbos species of the Cape Floristic Region of South Africa was not significantly improved following 30 minutes of aerosol smoke treatment (Brown and Botha 2004).

Haemodorum laxum **R. Br.**

BRANCHED BLOODROOT

Aerosol smoke treatments of broadcast seeds in rehabilitated mine sites in the southwest of Western Australia had no effect on the germination of this species (Roche et al. 1997a).

Haemodorum simplex **R. Br.**

SCENTED HAEMODORUM

The germination of *H. simplex*, which is widely distributed throughout the Southwest Botanical Province of Western Australia, was significantly enhanced when seeds were treated with aerosol smoke (Roche et al. 1997b).

Macropidia fuliginosa (Hook.) Druce

BLACK KANGAROO PAW

Dixon et al. (1995) reported that 90 minutes of cold smoke treatment had no effect on seed germination of this Western Australian plant species.

Wachendorfia paniculata L.

RED ROOT SEED AND ROOIKANOL (AFR.)

There were mixed results for this species of the Cape Floristic Region of South Africa. Seed germination was significantly improved following 30 minutes of aerosol smoke treatment, but seeds also germinated without smoke treatments (Brown and Botha 2004).

Wachendorfia thyrsiflora Burm.

BLOODROOT

Germination of bloodroot, a native to the Cape Floristic Region in South Africa, was promoted by aerosol smoke application (Brown et al. 2003; Brown and Botha 2004).

HYACINTHACEAE

Albuca flaccida Jacq.

SOLDIER-IN-A-BOX

Soldier-in-a-box occurs in the Cape Floristic Region of South Africa. The germination of its seeds was not promoted by aerosol smoke (Brown et al. 2003).

Albuca sp.

SENTRY BOXES

Seed germination in an unidentified species of *Albuca*, which also occurs in the Cape Floristic Region of South Africa, was not promoted by aerosol smoke (Brown et al. 2003).

IRIDACEAE

Aristea africana (L.) Hoffmanns

Brown et al. (2003) showed that germination of *A. africana*, which is native to the Cape Floristic Region in South Africa, doubled in response to aerosol smoke treatments (Brown and Botha 2004).

Aristea major Andrews

BLUE SCEPTER

Blue scepter occurs in the Cape Floristic Region of South Africa. Germination of its seeds was not significantly promoted by aerosol smoke (Brown et al. 2003).

Aristea racemosa **Baker.**

Aerosol smoke significantly increased seed germination in *A. racemosa*, a native to the Cape Floristic Region of South Africa, by 132% (Brown et al. 2003; Brown and Botha 2004).

Bobartia gladiata **(L. f.) Ker Gawl.**

Seed germination of this species, which occurs in the Cape Floristic Region of South Africa, was not significantly promoted by aerosol smoke (Brown et al. 2003).

Bobartia gladiata **ssp.** gladiata **(L. f.) Ker Gawl.**

Like *Bobartia gladiata*, which also occurs in the Cape Floristic Region of South Africa, germination in this subspecies was not significantly promoted by aerosol smoke (Brown et al. 2003).

Geissorhiza **sp.**

Seed germination in an unidentified species of *Geissorhiza*, commonly found in the Cape Floristic Region of South Africa, was not promoted by aerosol smoke (Brown et al. 2003).

Moraea ochroleuca **(Salisb.) Drapiez**

APRICOT TULP

Seed germination of this poisonous fynbos species of South Africa was not significantly improved after 30 minutes of exposure to aerosol smoke (Brown and Botha 2004).

Nivenia stokoei **(Guthrie) N. E. Br.**

STOKOE'S BUSH IRIS

Seed germination of this native South African fynbos plant was not significantly improved when exposed to 30 minutes of aerosol smoke (Brown and Botha 2004).

Orthrosanthus laxus **(Endl.) Benth.**

MORNING IRIS

Orthrosanthus laxus is a native of Western Australia. Its seeds germinated more readily when treated for 60 minutes with aerosol smoke (Roche et al. 1997b). Aerosol smoke treatments of broadcast seeds in rehabilitated mine sites in the southwest of Western Australia, in contrast, had no effect on their germination (Roche et al. 1997a). Norman et al. (2006) reported that neither aerosol smoke nor smoke water treatments significantly promoted germination of its seeds.

Patersonia babianoides **Benth.**

Aerosol smoke treatments of broadcast seeds in rehabilitated mine sites in the southwest of Western Australia had no effect on their germination (Roche et al. 1997a).

Patersonia glabrata **R. Br.**

LEAFY PURPLE-FLAG

Patersonia glabrata is native to the coastal heathlands and adjacent woodlands of eastern Australia. Sixty minutes exposure to aerosol smoke significantly improved seed germination in this species (Roche et al. 1997b).

Patersonia juncea **Lindl.**

RUSH-LEAVED PATERSONIA

Aerosol smoke treatments of broadcast seeds in rehabilitated mine sites in the southwest of Western Australia had no effect on their germination (Roche et al. 1997a).

Patersonia **sp.**

Soil samples from the Eden Burning Study Area, a dry sclerophyll forest in the Yalumba State Forest of New South Wales, Australia, were collected and air dried to test the effects of heat, smoke, and an interaction between the two cues on seeds from the seed bank. Samples exposed to heat treatment were incubated at 80°C for 60 minutes while those exposed to smoke were incubated in a room, where smoke was generated for 120 minutes. None of the treatments had any effect on the germination of the seeds of this species (Penman et al. 2008).

Patersonia **sp. aff.** fragilis

This species of *Patersonia* is native to southeastern Australia. Its seeds germinated more readily when exposed for 60 minutes to aerosol smoke (Roche et al. 1997b).

Patersonia occidentalis **R. Br.**

PURPLE FLAG

Purple flag is native to the proteaceous heathland, banksia woodland, and jarrah forest regions of Western Australia. The germination of this species significantly improved in response to smoke treatments (Dixon et al. 1995; Roche et al. 1997b). Roche et al. (1997a), Clarke and French (2005), and Norman et al. (2006) reported, in contrast, that neither aerosol smoke nor smoke water treatments significantly promoted germination of its seed.

Patersonia pygmaea **Lindl.**

PYGMY PATERSONIA

A 1 L/m^2 application of undiluted smoke water to the soil surface of a rehabilitated bauxite mine in the southwest of Western Australia had no significant effect on the germination of seeds from the soil seed bank (Roche et al. 1997a). An aerosol smoke treatment of broadcast seeds also did not promote germination.

Patersonia rudis **Endl.**

Aerosol smoke treatments of broadcast seeds in rehabilitated mine sites in the southwest of Western Australia had no effect on their germination (Roche et al. 1997a).

Patersonia umbrosa **Endl.**

SLENDER NATIVE IRIS

Like *P. rudis* above, aerosol smoke treatments of broadcast seeds in rehabilitated mine sites in the southwest of Western Australia had no effect on seed germination (Roche et al. 1997a).

Pillansia templemannii **L. Bolus**

This species occurs in the Cape Floristic Region of South Africa and in other parts of Africa. The germination of its seeds was not promoted by aerosol smoke (Brown et al. 2003).

Romulea **sp.**

Seed germination in an unidentified *Romulea* species, a native of South Africa, was not significantly improved after 30 minutes of exposure to aerosol smoke (Brown and Botha 2004).

Sisyrinchium **sp.**

DEVON SKIES

This species is native to North and South America, but has become a weed in many other parts of the world. In Australia, Read et al. (2000) reported that seed germination in this species was significantly improved with aerosol smoke treatments of 60 minutes.

Tritoniopsis parviflora **(Jacq.) G. J. Lewis**

This species occurs in the Cape Floristic Region of South Africa and elsewhere in Africa. The germination of its seeds was not promoted by aerosol smoke treatments (Brown et al. 2003).

Tritoniopsis triticea **(Burm. f.) Goldblatt**

Like *T. parvifora*, this species of *Tritoniopsis* also occurs in the Cape Floristic Region of South Africa. The germination of its seeds was not promoted by aerosol smoke (Brown et al. 2003).

Watsonia borbonica **(Pourr.) Goldblatt**

PINK WATSONIA

The pink watsonia occurs naturally in the Cape Floristic Region of South Africa and has naturalized in countries such as the United States and Australia. The germination of its seed was not promoted by aerosol smoke (Brown et al. 2003).

Watsonia tabularis **Mathews & L. Bolus**

TABLE MOUNTAIN WATSONIA

The Table Mountain watsonia occurs in the Cape Floristic Region of South Africa. Seed germination was not promoted by aerosol smoke (Brown et al. 2003).

JUNCACEAE

Juncus effusus **L. var.** decipiens **Buchenau**

LAMP RUSH

The effects of aerosol smoke, heat, darkness, cold stratification, and combinations of smoke with each of the three other treatments on seed germination were examined in this study (Tsuyuzaki and Miyoshi 2009). Smoke was produced by burning Timothy hay (*Phleum pratense*), which was pumped through a 3.5 m cooling tube into a smoke chamber for approximately 5 minutes. The seeds were exposed to the smoke for 60 minutes. Those seeds exposed also to heat were incubated at 75°C for 25 minutes. The cold stratification process took 1 month, during which the seeds remained in an incubator set to 4°C. Where the dark treatment was concerned, the seeds were maintained in total darkness for the entire germination period. Both the smoke and dark treatments significantly inhibited germination, with 45% and 0% germination, respectively compared to 97% germination for the control group. The seeds of this species, which occurs on most continents, display physiological dormancy.

Juncus planifolius **R. Br.**

BROAD LEAF RUSH

Soil samples from the Eden Burning Study Area, a dry sclerophyll forest in the Yalumba State Forest of New South Wales, Australia, were collected and air dried to test the effects of heat, smoke, and an interaction between the two cues on seeds from the seed bank. Samples exposed to heat treatment were incubated at 80°C for 60 minutes while those exposed to smoke were incubated in a room, where smoke was generated for 120 minutes. Only the smoke treatment significantly improved the germination of the seeds of this species (Penman et al. 2008). Neither heat nor the interaction between the two cues had any effect.

Luzula capitata **(Miq.) Nakai**

WOOD RUSH

The effects of aerosol smoke, heat, darkness, cold stratification, and combinations of smoke with each of the three other treatments on seed germination were examined in this study (Tsuyuzaki and Miyoshi 2009; see *Juncus effusus* var. *decipiens* above for details about the tests performed). None of the treatments had any effect on the germination of this species.

LANARIACEAE

Lanaria lanata **(L.) T. Durand & Schinz**

CAPE EDELWEISS

Cape edelweiss occurs in the Cape Floristic Region of South Africa. Seed germination in this species was not promoted by aerosol smoke (Brown et al. 2003).

PHORMIACEAE

Dianella amoena **G. W. Carr & P. F. Horfall**

MATTED FLAX-LILY

This Australian native plant occurs in the southeastern corner of Australia. Freshly collected seeds were exposed to 60 minutes of cool aerosol smoke without any significant improvement to germination (Roche et al. 1997b).

Dianella brevicaulis **(Ostenf.) G. W. Carr & P. F. Horsfall**

COAST FLAX LILY

Coast flax lily has a wide distribution throughout the southern region of the Southwest Botanic Province of Western Australia. Germination of *D. brevicaulis* increased significantly in response to aerosol smoke treatment. The final germination was 0% when given no pretreatment and 35.7% when seeds were treated with smoke (Roche et al. 1997b).

Dianella callicarpa **G. W. Carr & P. F. Horfall**

SWAMP FLAX-LILY

This lily occurs in the southern part of the state of Victoria, Australia. Freshly collected seeds were treated for 60 minutes with cool aerosol smoke, along with other treatments, but germination was not significantly promoted (Roche et al. 1997b).

Dianella **sp. aff.** longifolia **R. Br.**

BLUE FLAX-LILY

Blue Flax Lily is native to the tropical northern regions of Western Australia, Queensland, and Northern Territory, Australia. Germination was significantly improved when its seeds were treated with aerosol smoke for 60 minutes (Roche et al. 1997b).

Dianella revoluta **R. Br.**

BLUEBERRY LILY

Blueberry lily is a common component of the Southwest Botanical Province of Western Australia. Roche et al. (1997a,b) and Norman et al. (2006) reported that neither aerosol smoke nor smoke water treatments promoted germination of its seed.

Dianella revoluta **var. unknown R. Br.**

BLUEBERRY LILY

Blueberry lily occurs throughout the Southwest Botanical Province of Western Australia. Germination was significantly improved when treated for 60 minutes with cool aerosol smoke (Roche et al. 1997b).

Dianella tarda **G. W. Carr & P. F. Horsfall**

LATE-FLOWERED FLAX LILY

Dianella tarda is native to southeastern Australia. Germination was significantly improved when exposed to cool aerosol smoke for a 60 minute period (Roche et al. 1997b).

Dianella tasmanica **Hook. f.**

TASMANIAN FLAX LILY

This lily occurs in the states of Tasmania, Victoria, and New South Wales, Australia. Like some of the aforementioned flax-lilies, freshly collected seeds were treated for 60 minutes with cool aerosol smoke, some of which had received other treatments, but germination was not significantly promoted (Roche et al. 1997b).

Stypandra glauca **R. Br.**

NODDING BLUE LILY

This lily occurs in several Australian states. Freshly collected seeds were treated for 60 minutes with cool aerosol smoke, but germination was not significantly improved (Roche et al. 1997b).

POACEAE

Aira elegans **Willd. ex Kunth**

ANNUAL SILVER HAIRGRASS

Germination of annual silver hairgrass seeds, collected from a *Eucalyptus baxteri* heathy-woodland in Victoria, Australia, did not significantly improve following a 24-hour smoke water treatment (Enright and Kintrup 2001).

Agrostis scabra **Willd.**

ROUGH BENT GRASS

The effects of aerosol smoke, heat, darkness, cold stratification, and combinations of smoke with each of the three other treatments on seed germination were examined in this study (Tsuyuzaki and Miyoshi 2009). Smoke was produced by burning Timothy hay (*Phleum pratense*), which was pumped through a 3.5 m cooling tube into a smoke chamber for approximately 5 minutes. The seeds were exposed to the smoke for 60 minutes. Those seeds exposed also to heat were incubated at 75°C for 25 minutes. The cold stratification process took 1 month, during which the seeds remained in an incubator set to 4°C. Where the dark treatment was concerned, the seeds were maintained in total darkness for the entire germination period. Germination was significantly inhibited (17%) when the seeds of this species were treated with a combination of smoke and dark treatments. This was compared to 99% for the control group. None of the other treatments had any effect. This species is native to parts of Asia and North America.

Alopecurus myosuroides **Huds.**

BLACK GRASS

Alopecurus myosuroides is widely distributed across Europe, northern Africa and Asia. It has naturalized and become an invasive species in many other parts of the world. Germination almost doubled by treating its seeds with smoke water versus water only (Adkins and Peters 2001).

Amphipogon amphipogonoides **(Steud.) Vick.**

GREY-BEARDED GRASS

Roche et al. (1997a) reported that the germination of this native of Western Australian proteaceous heathlands, banksia woodlands, and jarrah forest communities doubled in response to smoke water application (1 L/m^2) to the soil seed bank of disturbed bauxite mine soils. A positive germination response was also observed when aerosol smoke was applied in situ to the jarrah forest soil (Roche et al. 1997a) and ex situ to seeds that had been sown prior to aerosol smoke treatment (Roche et al. 1997b). Norman et al. (2006) reported positive effects to aerosol smoke (60 minutes) for this species, especially when its seeds were incubated at an 18/10°C and light/dark 9/13 hour regime.

Andropogon ascinodis **C. B. Clarke.**

ANDROPOGAN

In this study, the seeds of *A. ascinodis* were soaked in smoke water (at concentrations of 100%, 75%, 50%, 25%, and 5% of the stock solution) for 24 hours. The smoke water stock solution was produced by burning a mixture of dominant native species from the Tiogo and Laba State forests of Sudan, and pumping the smoke into a bottle of water (volume not specified) for approximately 10 hours. The seeds were then shocked with heat, which consisted of incubating them for 2.5 minutes in a preheated kiln at 40, 80, 120, or 140°C. Germination capacity and mean germination time were recorded following the treatments (Dayamba et al. 2010). Neither smoke nor heat shock treatments had any effect on the germination of this species.

Andropogon gayanus **Kunth**

GAMBA GRASS

Dayamba et al. (2008) reported that a combination of aerosol smoke treatment and heat shock significantly inhibited seed germination in this fire-sensitive species. In another study by Dayamba et al. (2010), seeds collected from a Sudanian savanna-woodland in Burkina Faso, Africa, were treated with a variety of fire cues to also determine their effects on seed germination. The seeds were soaked in smoke water (at concentrations of 100%, 75%, 50%, 25%, and 5% of the stock solution) for 24 hours. The smoke water stock solution was produced by burning a mixture of dominant native species from the Tiogo and Laba State forests of Sudan, and pumping the smoke through water for 10 hours. The seeds also underwent a heat shock treatment, during which they were incubated in an oven at 40, 80, 120, or 140°C for 2.5 minutes. Following these treatments, germination percentages and mean germination times were measured. Where this species was concerned, neither the smoke nor heat shock treatments had any effect on the germination of this species.

Andropogon gerardii **Vitman**

BIG BLUE STEM

Big blue stem seed, collected from a tallgrass prairie in the Midwest of the United States, did not germinate in response to aerosol smoke treatments of 1, 10, or 60 minutes duration (Jefferson et al. 2007).

Aristida junciformis **Trin. & Rupr.**

GONGONI GRASS

Smoke water and karrikinolide (10^{-8} M) treatments, when applied at temperatures greater than 30°C, significantly increased the rate of germination and final germination of the seeds of this South African mesic grassland species (Ghebrehiwot et al. 2009).

Aristida ramosa **R. Br.**

PURPLE WIRE GRASS

Aristida ramosa occurs in Queensland, New South Wales, Victoria, and Western Australia. Germination of this species was inhibited by 22% when its seeds were treated with smoke water (10% dilution) (Clarke et al. 2000).

Aristida vagans **Cav.**

THREE-AWN SPEARGRASS

Three-awn speargrass is native to the dry sclerophyll forests of eastern New South Wales and Queensland, Australia, and has naturalized in New Zealand. Germination decreased by 5.75% in response to smoke water treatment, that is, soaking seeds in 10% smoke water for 24 hours (Clarke and French 2005).

Austrodanthonia caespitosa **(Gaudich.) H. P. Linder**

WHITE TOP

This grass is widely distributed across the southern states of Australia. Norman et al. (2006) showed that an aerosol smoke treatment of 60 minutes duration significantly improved the germination of white top, especially when its seeds were incubated in an 18/10°C and light/dark 9/13 hour regime.

Austrodanthonia racemosa **(R. Br.) H. P. Linder var.** racemosa

STRIPED WALLABY GRASS

Striped wallaby grass is native to eastern Australia and can be found growing in dry sclerophyll forests. Germination of this species was not significantly promoted when seeds were soaked in 10% smoke water for 24 hours (Clarke and French 2005).

Austrodanthonia tenuior **(Steud.) H. P. Linder**

WALLABY GRASS

Wallaby grass is also native to eastern Australia and occurs in dry sclerophyll forests. Germination of this species was promoted when its seeds were soaked in 10% smoke water for 24 hours (Clarke and French 2005).

Austrostipa compressa **(R. Br.) S. W. L. Jacobs & J. Everett**

Austrostipa compressa is a native of the southwestern region of Western Australia. Germination was improved when seeds were soaked in 5%, 10%, or 20% smoke water for 12 hours. Eighty percent germination occurred when seeds were treated compared to 20%–30% germination for the control (Smith et al. 1999). Baker et al. (2005) reported that smoke water treatments had no effect on seed germination, even when used in combination with other treatments.

Austrostipa macalpinei **(Reader) S. W. L. Jacobs & J. Everett**

ANNUAL SPEAR-GRASS

Baker et al. (2005) reported that smoke water treatments of seeds collected at Gin Gin, Western Australia, had no effect on the germination of this species, even when used in combination with other treatments, such as heat and manual scarification.

Austrostipa rudis **ssp.** rudis **(Spreng.) S. W. L. Jacobs & J. Everett**

VEINED SPEARGRASS

Veined speargrass commonly occurs in mixed eucalypt woodlands along the coastal regions and ranges of eastern Australia. Clarke and French (2005) showed that germination significantly increased as a result of soaking seeds in smoke water (10%) for 24 hours.

Austrostipa scabra **ssp.** falcata **(R. Br.) S. W. L. Jacobs & J. Everett**

FINE SPEARGRASS

Germination of fine speargrass, a native of Australia, was promoted by smoke treatments (Read and Bellairs 1999).

Austrostipa scabra **ssp.** scabra **(R. Br.) S. W. L. Jacobs & J. Everett**

ROUGH SPEARGRASS

Germination of rough speargrass, a native of Australia, was inhibited by smoke (Read and Bellairs 1999). Clarke et al. (2000) reported that smoke water had no effect on seed germination.

Avena fatua

Avena fatua **L.**

WILD OAT

This highly invasive species has naturalized throughout temperate regions of the world. The exact native range of *A. fatua* is not clear. In Australia, Adkins and Peters (2001) tested the effects of smoke water on seed germination of this species. Smoke water stimulated germination from 0% in the control to 92% (5%–20% smoke water) in 3-month old seed. The seeds of biotypes from western Canada, the northern United States, and England was also stimulated by smoke water. Smoke water inhibited germination when seeds were soaked in solution for 24 hours (Daws et al. 2007). A synthesized form based on naturally occurring karrikinolide (10^{-7} M), promoted seed germination in this species (Daws et al. 2007; Stevens et al. 2007).

Avena sterilis **ssp.** ludoviciana **L.**

ANIMATED OAT

Animated oat is indigenous to Europe, but has become invasive in many parts of the world. Adkins and Peters (2001) reported that germination was significantly promoted by smoke water (10% dilution), with 92% germination occurring following smoke treatment compared to 12% for the control group.

Bothriochloa macra **(Steud.) S. T. Blake**

RED GRASS

Bothriochloa macra occurs mainly in the grasslands of coastal, tableland and slope environments of the eastern states of Australia. Smoke water significantly promoted germination in this species (Clarke and French 2005). Earlier, Clarke et al.

(2000) reported that smoke water had no effect on red grass seeds collected from the New England bioregion between the years of 1995–1996.

Bouteloua curtipendula **(Michx.) Torr.**

SIDEOATS GRAMA

This species occurs throughout northern and southern America. Jefferson et al. (2007) reported that germination increased from 30% in control treatments to 80% when seeds were treated with aerosol smoke for 60 minutes. There was no significant improvement in germination when the seeds were exposed to aerosol smoke for 4 or 8 minutes, heat treatments of 30 or 60 seconds at 100°C, dry, cold stratification at 4°C for 1 month and relative humidity of 10% (Schwilk and Zavala 2012).

Bouteloua eriopoda **(Torr.) Torr.**

BLACK GRAMA

Aerosol smoke treatments of 4 or 8 minutes, in combination with heat treatments of 30 or 60 seconds at 100°C, as well as dry, cold stratification at 4°C for 1 month and relative humidity of 10%, had no effect on seed germination for this species (Schwilk and Zavala 2012). Black grama is native to the southwestern United States.

Bouteloua gracilis **(Willd. ex Kunth) Lag. ex. Griffiths**

BLUE GRAMA

Germination was significantly increased when the seeds were exposed to aerosol smoke treatments of 4 or 8 minutes, heat treatments of 30 or 60 seconds 100°C and were stratified for 1 month in a dry, cold environment at 4°C with a relative humidity of 10% (Schwilk and Zavala 2012). The increase was up to 100% following 4 minutes exposure to smoke and 80% following 8 minutes. Germination in the control group was 45%. This species is native to North America.

Brachiaria distichophylla **Stapf**

HAIRY SIGNAL GRASS

Seeds collected from a Sudanian savanna-woodland in Burkina Faso, Africa, were treated with a variety of fire cues to determine their effects on seed germination (Dayamba et al. 2010). The seeds were soaked in smoke water (at concentrations of 100%, 75%, 50%, 25%, and 5% of the stock solution) for 24 hours. The smoke water stock solution was produced by burning a mixture of dominant native species from the Tiogo and Laba State forests of Sudan, and pumping the smoke through water for 10 hours. The seeds also underwent a heat shock treatment, during which they were incubated in an oven at 40, 80, 120, or 140°C for 2.5 minutes. Following these treatments, germination percentages and mean germination times were measured. None of the smoke or heat shock treatments had any effect on the germination of this species.

Brachiaria lata (Schumach.) C. E. Hubb.

SIGNAL GRASS

Like *B. distichophylla*, the seeds were also collected from a Sudanian savanna-woodland in Burkina Faso, Africa, and treated with a variety of fire cues to determine their effects on seed germination (see *B. distichophylla* for details; Dayamba et al. 2010). None of the smoke or heat shock treatments had any effect on the germination of this species.

Bromus diandrus Roth

RIPGUT BROME

Ripgut brome is native to northern Africa, Europe and western Asia and is invasive in Australia and the United States. Adkins and Peters (2001) reported that germination increased significantly when seeds were treated with 5% smoke water. Stevens et al. (2007), however, reported that germination was inhibited when seeds were treated with a 1/10 (v/v) dilution of smoke water.

Bromus sterilis L.

POVERTY BROME

Poverty brome is a weed in many parts of the world including the United States, Australia and New Zealand. Its broad native range includes northern Africa, western Asia and Europe. Daws et al. (2007) showed that germination was inhibited when seeds were soaked for 24 hours in smoke water (Capeseed, Cape Town, South Africa).

Calamagrostis hakonensis (Franch. & Sav.) Keng.

The effects of aerosol smoke, heat, darkness, cold stratification, and combinations of smoke with each of the three other treatments on seed germination were examined in this study (Tsuyuzaki and Miyoshi 2009). Smoke was produced by burning Timothy hay (*Phleum pratense*), which was pumped through a 3.5 m cooling tube into a smoke chamber for approximately 5 minutes. The seeds were exposed to the smoke for 60 minutes. Those seeds exposed also to heat were incubated at 75°C for 25 minutes. The cold stratification process took 1 month, during which the seeds remained in an incubator set to 4°C. Where the dark treatment was concerned, the seeds were maintained in total darkness for the entire germination period. Germination was inhibited by smoke (34%) and dark (18%) treatments and promoted by the cold stratification treatment (88%). This was compared to 70% germination for the control group.

Chasmanthium latifolium (Michx.) Yates

INDIAN WOOD OATS

Indian wood oats seed, collected from a tallgrass prairie in the Midwest of the United States, did not germinate in response to aerosol smoke treatments of 1, 10, or 60 minutes duration (Jefferson et al. 2007).

Chloris ventricosa **R. Br.**

WINDMILL GRASS

Windmill grass is native to eastern Australia. Read and Bellairs (1999) showed that germination was significantly promoted by treating the seeds with smoke.

Cymbopogon refractus **(R. Br.) A. Camus**

BARBWIRE GRASS

Barbwire grass occurs on poor soils and is widely distributed throughout eastern and northern Australia. This species has become invasive in many parts of the world. Clarke and French (2005) reported that germination of this grass was unaffected by smoke water treatment unless the seeds had been shocked with heat at 40°C for 5 minutes. Seed germination was then significantly inhibited by smoke water.

Cymbopogon schoenanthus **Spreng.**

CAMEL GRASS

Seeds collected from a Sudanian savanna-woodland in Burkina Faso, Africa, were treated with a variety of fire cues to determine their effects on seed germination (Dayamba et al. 2010). The seeds were soaked in smoke water (at concentrations of 100%, 75%, 50%, 25%, and 5% of the stock solution) for 24 hours. The smoke water stock solution was produced by burning a mixture of dominant native species from the Tiogo and Laba State forests of Sudan, and pumping the smoke through water for 10 hours. The seeds also underwent a heat shock treatment, during which they were incubated in an oven at 40, 80, 120, or 140°C for 2.5 minutes. Following these treatments, germination percentages and mean germination times were measured. None of the smoke or heat shock treatments had any effect on the germination of this species.

Dactylis glomerata **L.**

ORCHARD GRASS

Orchard grass is native to the grasslands and open woodlands of Europe, northern Africa and parts of Asia. This species has become naturalized in other parts of the world. Pérez-Fernández and Rodríguez-Echeverría (2003) reported that *D. glomerata* germination was stimulated to over 90% when seeds were exposed to aerosol smoke for 20 minutes. This was compared to 50%–60% for the control group.

Danthonia caespitosa **Gaudich.**

WALLABY GRASS

Aerosol smoke treatments of broadcast seeds in rehabilitated mine sites in the southwest of Western Australia had no effect on the germination of wallaby grass (Roche et al. 1997a).

Dichanthium sericeum **(R. Br.) A. Camus**

QUEENSLAND BLUEGRASS

Dichanthium sericeum is widespread throughout the Australian continent and Papua New Guinea, but is more common to the tropical and subtropical regions of the world. Read and Bellairs (1999) reported that germination of this species was significantly promoted by smoke, while Clarke et al. (2000) and Clarke and French (2005) showed that smoke water had no significant effect.

Dichanthium setosum **S. T. Blake**

BLUEGRASS

Bluegrass is common to the grassy woodlands of the New England Tableland of New South Wales, Australia. Smoke water treatments did not significantly promote the germination of this species (Clarke et al. 2000).

Dichelachne micrantha **(Cav.) Domin**

SHORTHAIR PLUMEGRASS

This species occurs in some parts of the east and west coasts of Australia and is a weed in Hawaii, United States. Germination of its seeds was not significantly promoted when they were soaked in 10% smoke water for 24 hours (Clarke and French 2005).

Dichelachne rara **(R. Br.) Vick.**

COMMON PLUMEGRASS

Soil samples from the Eden Burning Study Area, a dry sclerophyll forest in the Yalumba State Forest of New South Wales, Australia, were collected and air dried to test the effects of heat, smoke, and an interaction between the two cues on seeds from the seed bank. Samples exposed to heat treatment were incubated at 80°C for 60 minutes while those exposed to smoke were incubated in a room, where smoke was generated for 120 minutes. Only the heat treatment significantly improved the germination of the seeds of this species (Penman et al. 2008). Neither smoke nor the interaction between the two cues had any effect.

Digitaria breviglumis **Henrard**

SHORT-GLUMED UMBRELLA GRASS

Digitaria breviglumis is native to the tropical eucalypt savanna and dry woodland communities of Queensland and New South Wales, Australia. Germination was promoted when the soil seed bank was exposed to aerosol smoke for 30 minutes (Williams et al. 2005).

Digitaria brownii **(Roem. & Schult.) Hughes**

COTTON PANIC GRASS

Cotton panic grass occurs throughout Australia. A 60 minute aerosol smoke treatment of soil samples containing seeds of this species, collected across forest edges between subtropical rainforest and eucalypt forest in the Lamington National Park of Queensland, Australia, did not significantly promote germination in the seeds of this species (Tang et al. 2003).

Digitaria ciliaris **(Retz.) Koeler**

SUMMER GRASS

Chou et al. (2012) tested the effects of smoke water, heat, and combinations of them both on the seeds of this species. Smoke treatments comprised of soaking seeds for 20 hours in the commercially available Regen 2000® smoke water solution, at concentrations of 1:5, 1:10, or 1:100 (v/v). The seeds were heated to 50 or 80°C for a period of 5 minutes. The heat shock treatment did not significantly affect germination percentage or mean germination time and there was no interaction between heat and smoke treatments. The germination percentage was, however, significantly inhibited and mean germination time significantly delayed by smoke water concentrations of 1:10 and 1:5.

Digitaria diffusa **Vick.**

OPEN-SUMMER GRASS

Digitaria diffusa is native to the open sclerophyll woodlands of Queensland and New South Wales, Australia. Germination of this species significantly increased in response to aerosol smoke treatment of 60 minutes (Read et al. 2000).

Digitaria ramularis **(Trin.) Henrard**

FINGERGRASS

Like *D. diffusa*, fingergrass is native to the open sclerophyll woodlands of Queensland and New South Wales, Australia. Germination of this species significantly increased in response to aerosol smoke treatment of 60 minutes (Read et al. 2000). Note that germination remained low regardless of seed pretreatment. Clarke and French (2005) reported, however, that germination was not significantly promoted when the seeds were soaked for 24 hours in 10% smoke water.

Echinopogon caespitosus **C. E. Hubb. var.** caespitosus

TUFTED HEDGEHOG GRASS

Tufted hedgehog grass occurs in the grasslands and mixed eucalypt woodlands of eastern New South Wales and southeastern Queensland, Australia. Soaking seeds in smoke water significantly inhibited germination (Clarke and French 2005).

Elymus hystrix **L. var.** hystrix

EASTERN BOTTLEBRUSH GRASS

Eastern bottlebrush grass seed, collected from a tallgrass prairie in the Midwest of the United States, did not germinate in response to aerosol smoke treatments of 1, 10, or 60 minutes duration (Jefferson et al. 2007).

Entolasia stricta **(R. Br.) Hughes**

WIRY PANIC

This species occurs on the east coast of Australia. Germination of its seed was not significantly promoted when they were soaked for 24 hours in 10% smoke water (Clarke and French 2005).

Eragrostis benthamii **Mattei**

BENTHAM'S LOVEGRASS

Eragrostis benthamii is native to the dry sclerophyll forests of southeastern Australia. Smoke water significantly enhanced seed germination in this species (Clarke and French 2005).

Eragrostis cilianensis **(All.) Janch.**

STINKGRASS

This native to Europe has naturalized and become weedy throughout the tropics. In Australia, Read et al. (2000) showed that germination was promoted when the soil seed bank had been treated for 60 minutes with aerosol smoke.

Eragrostis curvula **(Schrad.) Nees**

WEEPING LOVEGRASS

Weeping lovegrass is native to South Africa and is invasive in the United States and Australia. Clarke and French (2005) revealed that germination was modestly promoted by smoke water. The seeds were soaked in 10% smoke water for 24 hours. Smoke water treatments at 30°C significantly increased final germination percentage and shoot and root length, while treatment with karrikinolide (10^{-8} M) increased shoot and root length at various temperatures (Ghebrehiwot et al. 2009). Dormancy was also broken by seed storage for 5–6 months (Wasser 1982).

Eragrostis leptostachya **(R. Br.) Steud.**

AUSTRALIAN LOVEGRASS

Australian lovegrass grows in mixed eucalypt woodlands and grasslands in eastern Australia. Although this species responded positively to smoke water treatment (as described for *E. curvula*), germination was low (Clarke and French 2005). In an earlier study by Read et al. (2000), high germination (73%) was reached by treating the soil seed bank for 60 minutes with aerosol smoke. Heat treatment also stimulated germination to 83%.

Eragrostis sororia **Domin**

LOVEGRASS

This lovegrass occurs in the dry sclerophyll woodland communities of eastern Australia. Germination was enhanced by an aerosol smoke treatment of 60 minutes duration (Read et al. 2000).

Eragrostis tef (**Zucc.**) **Trotter**

TEFF LOVEGRASS

Smoke water and smoke-isolated karrikinolide significantly improved seed germination and seedling vigor under high temperature and low osmotic potential for this species of lovegrass (Ghebrehiwot et al. 2008).

Euclasta condylotricha **Stapf**

MOCK BLUESTEM

Seeds collected from a Sudanian savanna-woodland in Burkina Faso, Africa, were treated with a variety of fire cues to determine their effects on seed germination (Dayamba et al. 2010). The seeds were soaked in smoke water (at concentrations of 100%, 75%, 50%, 25%, and 5% of the stock solution) for 24 hours. The smoke water stock solution was produced by burning a mixture of dominant native species from the Tiogo and Laba State forests of Sudan, and pumping the smoke through water for 10 hours. The seeds also underwent a heat shock treatment, during which they were incubated in an oven at 40, 80, 120, or 140°C for 2.5 minutes. Following these treatments, germination percentages and mean germination times were measured. None of the treatments had any effect on germination.

Festuca megalura **Nutt.**

FOXTAIL FESCUE

Foxtail fescue, a native to Europe, has naturalized throughout the USA. Although germination was 85% without any seed pretreatment, heat shock and treatment with powdered charred wood increased germination to 96% (Keeley et al. 1985).

Festuca pallescens (**St.-Yves**) **Parodi**

THATCHING GRASS

The seeds of this native species of the grasslands of northwestern Patagonian were treated with a combination of heat (80°C for 5 minutes) and exposed to aerosol smoke for 60 minutes with no significant effect on germination (Gonzalez and Ghermandi 2012).

Heteropogon contortus (**L.**) **Roem. & Schult.**

BLACK SPEARGRASS

The native distribution of black speargrass is unknown. This grass is utilized as forage in Australia and South Africa and as a source of fiber in India. It has become invasive in many parts of the world. Germination of this species was promoted by smoke (Campbell 1995 cited in Read et al. 2000). Also known as piligrass, this native perennial bunchgrass is considered important to restoration and revegetation efforts in Hawaii. Baldos et al. (2011) attempted to establish methods for quickly breaking dormancy in the seeds of this species. Treating them with liquid smoke flavouring (1% v/v) significantly promoted germination (40.8%), and was more effective than using either gibberellic acid (10,000 ppm) or distilled water.

Heteropogon triticeus **(R. Br.) Stapf ex Craib**

GIANT SPEARGRASS

Heteropogon triticeus is a widespread tropical grass that occurs in northern Australia and Asia. Germination significantly increased when the soil seed bank of this species was exposed to aerosol smoke for 30 minutes (Williams et al. 2005).

Hordeum murinum **L. ssp.** leporinum **(Link) Arcang.**

BARLEY GRASS

Also synonymous with *Hordeum leporinum*, this weedy species is native to Europe, western Asia, the Caucasus and Soviet Middle Asia, and northern Africa and has naturalized in Australia, the United States, Korea, and New Zealand. Germination increased significantly in response to smoke water (10:10 v/v) and karrikinolide (0.67µM) treatments (Stevens et al. 2007). One hundred percent germination was achieved when treated with smoke water.

Hyparrhenia hirta **(L.) Staph**

COMMON THATCHING GRASS

Smoke water and karrikinolide (10^{-8} M), when applied at temperatures greater than 30°C, significantly increased rate of germination and final germination in the seeds of this South African mesic grassland species (Ghebrehiwot et al. 2009).

Imperata cylindrica **(L.) P. Beauv.**

BLADY GRASS

Blady grass is native to parts of Asia and Australia. A 60-minute aerosol smoke treatment of soil samples containing the seeds of this species, collected across forest edges between subtropical rainforest and eucalypt forest in the Lamington National Park of Queensland, Australia, did not significantly improve germination (Tang et al. 2003).

Loudetia togoensis **(Pilg.) C. E. Hubb.**

LOUDETIA

Seeds collected from a Sudanian savanna-woodland in Burkina Faso, Africa, were treated with a variety of fire cues to determine their effects on seed germination (Dayamba et al. 2010). The seeds were soaked in smoke water (at concentrations of 100%, 75%, 50%, 25%, and 5% of the stock solution) for 24 hours. The smoke water stock solution was produced by burning a mixture of dominant native species from the Tiogo and Laba State forests of Sudan, and pumping the smoke through water for 10 hours. The seeds also underwent a heat shock treatment, during which they were incubated in an oven at 40, 80, 120, or 140°C for 2.5 minutes. Following these treatments, germination percentages and mean germination times were measured. None of the treatments had any effect on the seeds of *L. togoensis*.

Microlaena stipoides **(Labill.) R. Br.**

WEEPING GRASS

Weeping grass is common to the grassy woodlands of the New England Tableland of New South Wales, Australia. Smoke water treatments had no effect on the germination, positive or negative, of this species, which usually germinates to 100% (Clarke et al. 2000). In another study, soil samples from the Eden Burning Study Area, a dry sclerophyll forest in the Yalumba State Forest of New South Wales, Australia, were collected and air dried to test the effects of heat, smoke, and an interaction between the two cues on seeds from the seed bank (Penman et al. 2008). Samples exposed to heat treatment were incubated at 80°C for 60 minutes while those exposed to smoke were incubated in a room, where smoke was generated for 120 minutes. Both the heat treatment and an interaction between heat and smoke induced a marginally significant increase in the germination of the seeds of this species. The smoke treatment alone had no effect.

Microlaena stipoides **(Labill.) R. Br. var.** stipoides

WEEPING GRASS

This species occurs in several places in Australia. Germination of its seeds was not significantly promoted when they were soaked for 24 hours in 10% smoke water (Clarke and French 2005).

Miscanthus sinensis **Andersson**

CHINESE SILVERGRASS

The effects of aerosol smoke, heat, darkness, cold stratification, and combinations of smoke with each of the three other treatments on seed germination were examined in this study (Tsuyuzaki and Miyoshi 2009). Smoke was produced by burning Timothy hay (*Phleum pratense*), which was pumped through a 3.5 m cooling tube into a smoke chamber for approximately 5 minutes. The seeds were exposed to the smoke for 60 minutes. Those seeds exposed also to heat were incubated at 75°C for 25 minutes. The cold stratification process took 1 month, during which the seeds remained in an incubator set to 4°C. Where the dark treatment was concerned, the seeds were maintained in total darkness for the entire germination period. The smoke treatment inhibited germination (76%). This was compared to 95% germination for the control group. None of the other treatments had any effect on germination. This species is common to parts of eastern Asia, including China and Japan.

Neurachne alopecuroidea **R. Br.**

FOXTAIL MULGA GRASS

Foxtail mulga grass is a native of southwestern Western Australia's proteaceous heathland, banksia woodland, and jarrah forest regions. Ex situ germination significantly improved when the soil seed bank was exposed to aerosol smoke for 90 minutes (Dixon et al. 1995). Norman et al. (2006) reported, in contrast, that neither aerosol smoke nor smoke water treatments significantly promoted germination of its seed.

Notodanthonia racemosa **(R. Br.) Zotov**

WALLABY GRASS

Wallaby grass is common to the grassy woodlands of the New England Tableland of New South Wales, Australia. Smoke water treatments did not significantly promote germination in this species (Clarke et al. 2000).

Notodanthonia richardsonii **(Cashmore) Veldkamp**

WALLABY GRASS

Like *N. racemosa* above, this species of wallaby grass is also common to the grassy woodlands of the New England Tableland of New South Wales, Australia. Smoke water treatments did not significantly promote germination of its seeds either (Clarke et al. 2000).

Oplismenus aemulus **(R. Br.) Roem. & Schult.**

BASKET GRASS

This tussock grass occurs in the subtropical and temperate regions of eastern Australia. Aerosol smoke applications of 60 minutes duration to the soil seed bank promoted seed germination (Tang et al. 2003).

Oryza sativa **L.**

RICE

Rice is extensively cultivated throughout the tropic, subtropic, and warm-temperate regions of the world. Kulkarni et al. (2006) reported that smoke water (1:500 dilution) and karrikinolide (10^{-8} to 10^{-10} M) promoted root and shoot elongation, as well as the number of lateral roots. In addition, the seedling vigor index was higher than that of untreated seeds. Doherty and Cohn (2000) reported that a commercial liquid smoke flavoring product broke dormancy in intact and dehulled red rice.

Panicum decompositum **R. Br.**

NATIVE MILLET

Native millet is widespread throughout the states of New South Wales, Queensland, South Australia and Victoria, Australia. Read and Bellairs (1999) showed that ex situ germination increased from 7.7% with no treatment to 63% when seeds were treated with smoke.

Panicum effusum **R. Br.**

HAIRY PANIC GRASS

Hairy panic grass has a widespread distribution in woodland and disturbed sites throughout Australia. Read et al. (2000) were able to achieve 14% germination when they exposed the soil seed bank for 60 minutes to aerosol smoke. Clarke and French (2005) reported there was 10% germination when the seeds were soaked for 24 hours in 10% smoke water.

Panicum maximum **Jacq.**

GUINEA GRASS

Smoke water and karrikinolide (10^{-8} M) treatment, when applied at temperatures greater than 30°C, significantly increased rate of germination and final germination percentage in the seeds of this South African mesic grassland species (Ghebrehiwot et al. 2009).

Panicum simile **Domin**

TWO-COLOUR PANIC

Two-colour panic grows in mixed eucalypt woodlands and scrub in eastern Australia. Soaking the seeds in smoke water (10% dilution) for 24 hours improved germination in this species (Clarke and French 2005).

Panicum virgatum **L.**

SWITCH GRASS

Chou et al. (2012) tested the effects of smoke water, heat and combinations of them both on the seeds of this species. Smoke treatments comprised of soaking seeds for 20 hours in the commercially available Regen 2000® smoke water solution, at concentrations of 1:5, 1:10, or 1:100 (v/v). The seeds were then heated to 50 or 80°C for a period of 5 minutes. Only the smoke water treatment of 1:5 produced any response by significantly inhibiting germination percentage. Mean germination time was not affected and there was no response due to heat or interactions with it.

Paspalidium distans **(Trin.) Hughes**

SPREADING PANIC GRASS

Spreading panic grass is native to northern and eastern Australia, as well as Papua New Guinea, and has become a weed in Hawaii. Read et al. (2000) showed that germination was promoted when seeds were treated for 60 minutes with an aerosol smoke treatment. Soaking the seeds in a 10% smoke water solution for 24 hours prior to sowing had no significant effect on germination (Clarke and French 2005).

Pentaschistis colorata **(Steud.) Stapf**

This native of the Cape Floristic Region of South Africa would not germinate without first being exposed to aerosol (Brown et al. 2003; Brown and Botha 2004).

Phalaris paradoxa **L.**

HOOD CANARYGRASS

Hood canarygrass is native to northern Africa, Europe, and western Asia and is invasive to Australia and the United States. Adkins and Peters (2001) reported that germination increased significantly when its seeds were treated with a 5% smoke water solution.

Poa labillardieri **Steud.**

COMMON TUSSOCK GRASS

Common tussock grass is widely distributed along the coastal regions of Queensland, New South Wales, Victoria, Tasmania, and South Australia. Germination was enhanced by smoke application (Read and Bellairs 1999).

Poa sieberiana **Spreng.**

GREY TUSSOCK GRASS

Grey tussock is common to the grassy woodlands of the New England Tableland of New South Wales, Australia. Smoke water treatments had no effect on the germination of this species, which usually germinates to 100% (Clarke et al. 2000).

Pseudopentameris macrantha **(Schrad.) Conert**

Germination of *P. macrantha*, a native of the Cape Floristic Region of South Africa, significantly increased in response to aerosol smoke treatment (Brown et al. 2003; Brown and Botha 2004).

Rottboellia exaltata **L. f.**

ITCH GRASS

Seeds collected from a Sudanian savanna-woodland in Burkina Faso, Africa, were treated with a variety of fire cues to determine their effects on seed germination (Dayamba et al. 2010). The seeds were soaked in smoke water (at concentrations of 100%, 75%, 50%, 25%, and 5% of the stock solution) for 24 hours. The smoke water stock solution was produced by burning a mixture of dominant native species from the Tiogo and Laba State forests of Sudan, and pumping the smoke through water for 10 hours. The seeds also underwent a heat shock treatment, during which they were incubated in an oven at 40, 80, 120, or 140°C for 2.5 minutes. Following these treatments, germination percentages and mean germination times were measured. None of the s smoke or heat shock treatments had a significant effect on the germination of this species.

Schizachyrium scoparium **(Michx.) Nash var.** scoparium

LITTLE BLUE STEM

Little blue stem seed, collected from a tallgrass prairie in the Midwest of the United States, did not germinate in response to aerosol smoke treatments of 1, 10, or 60 minutes duration (Jefferson et al. 2007).

Sorghastrum nutans **(L.) Nash**

INDIAN GRASS

Indian grass seed, like little blue stem, and also collected from a tallgrass prairie in the Midwest of the United States, did not germinate in response to aerosol smoke treatments of 1, 10, or 60 minutes (Jefferson et al. 2007).

Sorghum halepense **(L.) Pers.**

JOHNSON GRASS

Although the exact native range of this species is not known (most likely northern Africa, Western Asia, India, and Pakistan), Johnson grass has become invasive of the warm-temperate regions of the world. An Australian study by Adkins and Peters (2001) revealed that 100% germination was achieved by treating the seeds with 50% smoke water. Twelve percent germination occurred in the control (water only). Germination was also promoted when seeds were treated with karrikinolide (10^{-7} M; Daws et al. 2007).

Sorghum leiocladum **(Hack.) C. E. Hubb.**

WILD SORGHUM

Wild sorghum has a widespread distribution across northern and eastern Australia. Smoke water (10%) doubled the germination of this species. Germination in the dark or heat shock (80°C for 15 minutes), however, promoted greater germination (Clarke et al. 2000).

Sporobolus heterolepis **(A. Gray) A. Gray**

PRAIRIE DROPSEED

Sporobolus heterolepis seed, collected from a tallgrass prairie in the Midwest of the United States, did not germinate in response to aerosol smoke treatments of 1, 10, or 60 minutes duration (Jefferson et al. 2007).

Sporobolus indicus **(L.) R. Br. var.** capensis **Engelm.**

AFRICAN DROPSEED

Clarke and French (2005) reported that the germination of African dropseed was inhibited by smoke water treatment.

Tetrarrhena juncea **R. Br.**

FOREST WIRE GRASS

Soil samples from the Eden Burning Study Area, a dry sclerophyll forest in the Ya-lumba State Forest of New South Wales, Australia, were collected and air dried to test the effects of heat, smoke, and an interaction between the two cues on seeds from the seed bank. Samples exposed to heat treatment were incubated at 80°C for 60 minutes while those exposed to smoke were incubated in a room, where smoke was generated for 120 minutes. Both the smoke treatment and an interaction between heat and smoke induced a marginally significant increase in the germination of the seeds of this species (Penman et al. 2008). The heat treatment alone had no effect.

Tetrarrhena laevis **R. Br.**

FOREST RICE GRASS

Forest rice grass occurs in the proteaceous heathland, banksia woodland, and jarrah forest communities of the southwest region of Western Australia. Exposure to aerosol smoke for 60 minutes significantly promoted germination of seeds sown on the disturbed areas of a bauxite mine in Western Australia (Roche et al. 1997a). Similar results were achieved when the treated seeds were sown in punnets of soil and germinated in a glasshouse (Roche et al. 1997b), but smoke water had no effect (Roche et al. 1997a). Norman et al. (2006) reported that neither aerosol smoke nor smoke water treatments promoted germination of forest rice grass seeds.

Themeda australis **(R. Br.) Stapf**

KANGAROO GRASS

This species is a perennial grass that once occurred throughout Australia. Germi-nation of its seeds was not significantly promoted when they were soaked for 24 hours in 10% smoke water (Clarke and French 2005).

Themeda triandra **Forssk.**

KANGAROO GRASS

Themeda triandra is native to the South African mesic grasslands and savanna eco-systems of Africa, Asia and Australia. The seeds of this species responded positively to aerosol smoke treatments (Baxter et al. 1994; Brown et al. 1995; Brown et al. 2003; Brown and Botha 2004), to smoke water, and karrikinolide (10^{-8} M), with both rate of germination and final germination increasing significantly (Ghebrehiwot et al. 2009). Clarke et al. (2000) reported that smoke water had no effect on seeds collected from the grassy woodlands of the New England Tableland of New South Wales, Australia.

Triodia longiceps **J. M. Black**

GIANT GREY SPINIFEX

Triodia longiceps is common in the semiarid and arid regions of Australia. David-son and Adkins (1997) reported that plant-derived smoke promoted the germina-tion of this species (cited in Read et al. 2000).

Tristachya leucothrix **Trin. ex Nees**

TRIDENT GRASS

Smoke water and karrikinolide (10^{-8} M) significantly increased rate of germination and final germination in the seeds of this South African mesic grassland species (Ghebrehiwot et al. 2009).

Zea mays **L.**

MAIZE

Sparg et al. (2006) have suggested that smoke has the potential to improve germination and seedling vigor in the commercial maize cultivar, *Zea mays* L. var. PAN 6479. Prolonged exposure to aerosol smoke inhibited germination but was reversed by rinsing the seeds. Modi (2002, 2004) reported that traditional smoke-over-fire storage of native subsistence farmers in South Africa helped improve seed quality, germination, and seedling vigor of at least two landraces of maize.

RESTIONACEAE

Askidiosperma andreaeanum **(Pillans) H. P. Linder**

Germination of this native to the Cape Floristic Region of South Africa significantly increased in response to aerosol smoke treatment (Brown et al. 1995; Brown et al. 2003; Brown and Botha 2004).

Askidiosperma chartaceum **(Pillans) H. P. Linder**

Like *A. andreaeanum*, *A. chartaceum* also occurs in the Cape Floristic Region of South Africa. Seed germination in this species was not, however, promoted by aerosol smoke (Brown et al. 2003).

Askidiosperma esterhuyseniae **(Pillans) H. P. Linder**

Like other *Askidiosperma* species above, this species also occurs in the Cape Floristic Region of South Africa. Seed germination in this species was not significantly improved by exposure to aerosol smoke (Brown et al. 2003).

Askidiosperma paniculatum **(Mast.) H. P. Linder**

This species also occurs in the Cape Floristic Region of South Africa. Seed germination in this species was not promoted by aerosol smoke (Brown et al. 2003).

Baloskion tetraphyllum **(Labill.) B. G. Briggs & L. A. Johnson.**

PLUME RUSH

The active substance in plant-derived smoke, karrikinolide, can, in addition to promoting seed germination in *Baloskion tetraphyllum* (Syn. *Restio tetraphyllus*), stimulate the development and maturation of somatic embryos in this species (Ma et al. 2007).

Calopsis impolita **(Kunth) H. P. Linder**

This species occurs in the Cape Floristic Region of South Africa. Seed germination was not promoted by treatment with aerosol smoke (Brown et al. 2003).

Calopsis paniculata **Desv.**

KANET

Kanet also occurs in the Cape Floristic Region of South Africa. Seed germination was not promoted by aerosol smoke in this species either (Brown et al. 2003).

Cannomois parviflora **(Thunb.) Pillans**

Seed germination of this native to the Cape Floristic Region of South Africa was not significantly improved following 30 minutes of aerosol smoke treatment (Brown et al. 2003; Brown and Botha 2004).

Cannomois taylorii **H. P. Linder**

Like *C. parviflora*, seed germination of this native to the Cape Floristic Region of South Africa was not significantly improved following 30 minutes of aerosol smoke treatment (Brown and Botha 2004).

Cannomois virgata **(Rottb.) Steud.**

BELLREED

Germination of this native to the Cape Floristic Region of South Africa significantly increased in response to aerosol smoke treatment (Brown et al. 1995; Brown et al. 2003; Brown and Botha 2004).

Ceratocaryum argenteum **Nees ex Kunth**

ARROW REEDS

Seed germination of this native to the Cape Floristic Region of South Africa was not significantly improved in response to 30 minutes of aerosol smoke treatment (Brown and Botha 2004).

Chondropetalum ebracteatum **(Kunth) Pillans**

LARGE MILLET REED

Germination of this native to the Cape Floristic Region of South Africa was significantly improved following aerosol smoke treatment (Brown et al. 1995; Brown et al. 2003; Brown and Botha 2004).

Chondropetalum hookerianum **(Mast.) Pillans**

MEDIUM MILLET REED

Germination of this native to the Cape Floristic Region of South Africa significantly increased in response to aerosol smoke treatment (Brown et al. 1995; Brown et al. 2003; Brown and Botha 2004).

Chondropetalum mucronatum **(Nees) Pillans**

GIANT MILLET REED

Seed germination of this South African fynbos species increased from 4% without treatment to 81% when treated with aerosol smoke (Brown et al. 1995; Brown et al. 2003; Brown and Botha 2004).

Chondropetalum tectorum **(L. f.) Raf.**

Seed germination in this native to the Cape Floristic Region of South Africa significantly increased in response to aerosol smoke treatment (Brown et al. 1995; Brown et al. 2003; Brown and Botha 2004).

Dovea macrocarpa **Kunth**

Germination of this native to the Cape Floristic Region of South Africa significantly increased in response to aerosol smoke treatment (Brown et al. 1995; Brown et al. 2003; Brown and Botha 2004).

Elegia caespitosa **Esterh.**

Seed germination of this native to the Cape Floristic Region of South Africa was not significantly improved in response to 30 minutes of aerosol smoke treatment (Brown and Botha 2004). The authors note, however, that the seeds tested were probably of low viability.

Elegia capensis **(Burm. f.) Schelpe**

BROOM REED

The germination of this South African fynbos species was increased by 333% when seeds were treated for 30 minutes with aerosol smoke (Brown et al. 1995; see also Brown and Botha 2004). In a subsequent study, there was no significant effect on germination following smoke treatment (Brown et al. 2003).

Elegia cuspidata **Mast.**

BLOMBIESIE (AFR.)

Germination of this native of the Cape Floristic Region of South Africa significantly increased in response to aerosol smoke treatment (Brown et al. 1995; Brown et al. 2003; Brown and Botha 2004).

Elegia equisetacea **(Mast.) Mast.**

Seed germination in this native of South Africa was significantly increased after treatment with aerosol smoke (Brown et al. 1995; Brown et al. 2003; Brown and Botha 2004).

Elegia fenestrata **Pillans**

Germination of this native to the Cape Floristic Region of South Africa significantly increased in response to aerosol smoke treatment (Brown et al. 1995;

Brown et al. 2003; Brown and Botha 2004). This species is considered vulnerable (IUCN 2013).

Elegia filacea **Mast.**

LITTLE GOLDEN CURLS

Like other *Elegia* species above, seed germination in this native to the Cape Floristic Region of South Africa significantly increased in response to aerosol smoke treatment (Brown et al. 1995; Brown et al. 2003; Brown and Botha 2004).

Elegia grandis **(Nees) Kunth**

Germination in this species, which is native to the Cape Floristic Region of South Africa, significantly increased following aerosol smoke treatment (Brown et al. 1995; Brown et al. 2003; Brown and Botha 2004).

Elegia grandispicata **H. P. Linder**

This species also occurs in the Cape Floristic Region of South Africa. Aerosol smoke did not improve seed germination (Brown et al. 2003).

Elegia persistens **Mast.**

The germination of the seeds of this native of the Cape Floristic Region of South Africa was significantly improved with aerosol smoke treatments (Brown et al. 1995; Brown et al. 2003; Brown and Botha 2004).

Elegia spathacea **Mast.**

Germination of this native to the Cape Floristic Region of South Africa significantly increased in response to aerosol smoke treatment (Brown et al. 1995; Brown et al. 2003; Brown and Botha 2004).

Elegia stipularis **Mast.**

CUSHION RESTIO

Cushion restio occurs in the Cape Floristic Region of South Africa. Aerosol smoke did not improve seed germination (Brown et al. 2003).

Elegia thyrsifera **(Rottb.) Pers.**

Like *E. stipularis*, this is one of many species of plants that occurs in the Cape Floristic Region of South Africa and whose germination was not affected by aerosol smoke treatments (Brown et al. 2003).

Hydrophilus rattrayi **(Pillans) H. P. Linder**

This species occurs in the Cape Floristic Region of South Africa. An application of aerosol smoke did not improve its seed germination (Brown et al. 2003).

Hypodiscus neesii **Mast.**

Seed germination of this native fynbos species of the Cape Floristic Region of South Africa was not significantly improved following aerosol smoke treatments (Brown et al. 2003; Brown and Botha 2004).

Hypodiscus **sp.**

Germination of this native to the Cape Floristic Region of South Africa significantly increased in response to aerosol smoke treatment (Brown et al. 2003; Brown and Botha 2004).

Hypodiscus striatus **(Kunth) Mast.**

CAPE REED

Like *H. neesii*, seed germination of this native fynbos species of the Cape Floristic Region of South Africa was not significantly improved following aerosol smoke treatments (Brown et al. 2003; Brown and Botha 2004).

Ischyrolepis ocreata **(Kunth) H. P. Linder**

Germination of this native species of the Cape Floristic Region of South Africa significantly increased after treatment with aerosol smoke (Brown et al. 2003; Brown and Botha 2004).

Ischyrolepis sieberi **(Kunth) H. P. Linder**

Seed germination of this South African fynbos species increased in response to smoke water treatment. A dilution of 1:10 yielded the highest final germination percentages (92%) (Brown 1993a). Brown et al. (1995) achieved 69% germination using aerosol smoke in comparison to 1% for the control (see also Brown et al. 2003, Brown and Botha 2004).

Ischyrolepis subverticillata **Steud.**

Seed germination in this plant species of the Cape Floristic Region of South Africa significantly increased in response to aerosol smoke treatment (Brown et al. 2003; Brown and Botha 2004).

Mastersiella digitata **(Thunb.) Gilg-Ben.**

Thirty minutes exposure to aerosol smoke did not significantly improve seed germination in this native fynbos species of the Cape Floristic Region of South Africa (Brown et al. 2003; Brown and Botha 2004).

Restio bifarius **Mast.**

Germination of this native to the Cape Floristic Region of South Africa significantly increased in response to aerosol smoke treatment (Brown et al. 2003; see also Brown and Botha 2004).

Restio brachiatus **(Mast.) Pillans**

Brown et al. (1995) reported that germination of *R. brachiatus*, a native of South Africa, fynbos region, increased by 143% when its seeds were treated for 30 minutes with aerosol smoke. A subsequent study resulted in no significant improvement in seed germination (Brown et al. 2003).

Restio dispar **Mast.**

Seed germination in this native of the fynbos region of South Africa was significantly improved with an aerosol smoke treatment (Brown et al. 2003; Brown and Botha 2004).

Restio festuciformis **Nees ex Mast.**

GREEN GRASS REED

Seed germination in this native of the fynbos region of South Africa was significantly improved with an aerosol smoke treatment (Brown et al. 2003; Brown and Botha 2004). This species has been classified as vulnerable (IUCN 2013).

Restio pachystachyus **Kunth**

This species also occurs in the Cape Floristic Region of South Africa. An application of aerosol smoke did not significantly improve seed germination (Brown et al. 2003).

Restio praeacutus **Mast.**

Restio praeacutus is a native South African fynbos species. Germination was significantly improved following an aerosol smoke seed treatment of 60 minutes (Brown et al. 1995). A subsequent study resulted in no significant improvement in seed germination (Brown et al. 2003).

Restio similis **Pillans**

SLENDER BLOBS

Smoke water applications to the seeds of this South African fynbos species inhibited germination at a dilution of 1:10 and promoted germination at a dilution of 1:50 (Brown 1993a). Aerosol smoke also stimulated germination of this species (Brown 1993b; Brown et al. 1995; Brown et al. 2003).

Restio tetragonus **Thunb.**

Seed germination in this native of the fynbos region of South Africa was significantly improved with an aerosol smoke treatment (Brown et al. 2003; Brown and Botha 2004). Two percent germination was achieved without treatment compared to 97% with treatment.

Restio triticeus **Rottb.**

Seed germination in this native of the fynbos region of South Africa was significantly improved with an aerosol smoke treatment (Brown et al. 2003; Brown and

Botha 2004). A total of 37% germination was achieved without treatment compared to 94% with treatment.

Rhodocoma arida **H. P. Linder & Vlok**

Seed germination in this native of the fynbos region of South Africa was significantly improved with an aerosol smoke treatment (Brown et al. 2003; Brown and Botha 2004). Flematti et al. (2004) showed that the germination of this species was also promoted in response to karrikinolide treatments (10 ppb).

Rhodocoma capensis **Steud.**

CAPE RHODOCOMA

Seed germination in this native of the fynbos region of South Africa was significantly improved with an aerosol smoke treatment (Brown et al. 2003; Brown and Botha 2004).

Rhodocoma fruticosa **(Thunb.) H. P. Linder**

Seed germination in this native of the fynbos region of South Africa was significantly improved with an aerosol smoke treatment (Brown et al. 2003; Brown and Botha 2004).

Rhodocoma gigantea **(Kunth) H. P. Linder**

GIANT RHODOCOMA

Seed germination in this native of the fynbos region of South Africa was significantly improved with an aerosol smoke treatment (Brown et al. 2003; Brown and Botha 2004).

Staberoha aemula **(Kunth) Pillans**

Seed germination in this native of the fynbos region of South Africa was significantly improved with an aerosol smoke treatment (Brown et al. 2003; Brown and Botha 2004).

Staberoha banksii **Pillans**

Seed germination in this native of the fynbos region of South Africa was significantly improved with an aerosol smoke treatment (Brown et al. 2003; Brown and Botha 2004).

Staberoha cernua **(L. f.) T. Durand & Schinz**

Seed germination in this native of the fynbos region of South Africa was significantly improved with an aerosol smoke treatment (Brown et al. 2003; Brown and Botha 2004).

Staberoha distachyos **(Rottb.) Kunth**

Seed germination in this native of the fynbos region of South Africa was significantly improved with an aerosol smoke treatment (Brown et al. 2003; Brown and Botha 2004).

Staberoha disticha T. Durand & Schinz

Germination of *S. disticha*, a South African fynbos species, significantly increased in response to 30 minutes exposure to aerosol smoke (Brown 1993a).

Staberoha vaginata (Thunb.) Pillans

Seed germination in this native of the fynbos region of South Africa was significantly improved following an aerosol smoke treatment (Brown et al. 2003; Brown and Botha 2004). Note that germination was low regardless of treatment.

Thamnochortus bachmannii Mast.

STEENBOK REED

Seed germination in this native of the fynbos region of South Africa was significantly improved with an aerosol smoke treatment (Brown et al. 2003; Brown and Botha 2004).

Thamnochortus cinereus H. P. Linder

SILVER REED

Seed germination in this native of the fynbos region of South Africa was significantly improved with an aerosol smoke treatment (Brown et al. 2003; Brown and Botha 2004).

Thamnochortus erectus (Thunb.) Mast.

This species occurs in the Cape Floristic Region of South Africa. An application of aerosol smoke did not improve its seed germination (Brown et al. 2003).

Thamnochortus insignis Mast.

THATCHING REED

Seed germination in this native of the fynbos region of South Africa was significantly improved with an aerosol smoke treatment (Brown et al. 2003; Brown and Botha 2004).

Thamnochortus lucens (Poir.) H. P. Linder

This species occurs in the Cape Floristic Region of South Africa. An application of aerosol smoke did not improve its seed germination (Brown et al. 2003).

Thamnochortus pellucidus Pillans

Germination of this South African fynbos species was significantly increased in response to a 30-minute exposure to aerosol smoke (Brown 1993a). This species is considered vulnerable IUCN (2013).

Thamnochortus platypteris Kunth

Germination of this South African fynbos species increased in response to aerosol smoke, but was low (Brown and Botha 2004). A previous study reported no significant improvement to seed germination (Brown et al. 2003).

Thamnochortus rigidus **Esterh.**

This species occurs in the Cape Floristic Region of South Africa. An application of aerosol smoke did not significantly improve seed germination (Brown et al. 2003).

Thamnochortus punctatus **Pillans**

SANDVELD THATCHING REED

Germination of this native to the Cape Floristic Region of South Africa significantly increased in response to aerosol smoke treatment (Brown et al. 2003; see also Brown and Botha 2004).

Thamnochortus spicigerus **(Thunb.) Spreng.**

DUNE REED

Germination of this native to the Cape Floristic Region of South Africa significantly increased in response to aerosol smoke treatment (Brown *et al.* 2003; see also Brown and Botha 2004).

Thamnochortus sporadicus **Pillans**

Aerosol smoke promoted the germination of this South African fynbos species (Brown et al. 1995; Brown et al. 2003; Brown and Botha 2004).

Willdenowia incurvata **(Thunb.) H. P. Linder**

Germination of *W. incurvata*, a native of South Africa, responded positively to aerosol smoke treatment (Brown and Botha 2004). Germination was, however, still less than 10% for smoke-treated seeds. A previous study revealed no significant germination following aerosol smoke treatment (Brown et al. 2003).

SMILACACEAE

Smilax australis **R. Br.**

LAWYER VINE

Lawyer vine is endemic to Australia. A 60 minute aerosol smoke treatment of soil samples containing the seeds of this species, collected across forest edges between subtropical rainforest and eucalypt forest in the Lamington National Park of Queensland, Australia, did not germinate (Tang et al. 2003).

XANTHORRHOEACEAE

Xanthorrhoea johnsonii **A. T. Lee**

JOHNSON'S GRASSTREE

Johnson's grasstree is common to the grassy woodlands of the New England Tableland of New South Wales, Australia. Smoke water treatments had no effect on the germination of this species (Clarke et al. 2000).

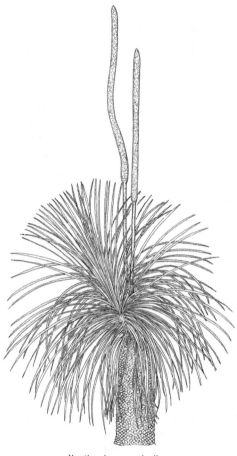

Xanthorrhoea preissii

Xanthorrhoea **sp.**

GRASSTREES

An unidentified species of *Xanthor-rhoea* occurs in the Southwest Botanical Province of Western Australia. Roche et al. (1997a) showed that the seeds of the species were not promoted by applying smoke water to the soil seed bank of a bauxite mine site undergoing rehabilitation. Aerosol smoke did, however, significantly promote germination in broadcast seed.

Dicotyledonae

ACANTHACEAE

Pseudoeranthemum variabile **(R. Br.) Radlk.**

LOVE FLOWER

Love flower is native to eastern Australia. Soil samples containing the seeds of this species, collected across forest edges between subtropical rainforest and eucalypt forest in the Lamington National Park of Queensland, Australia, was treated for 60 minutes with cool aerosol smoke. There was no improvement in germination (Tang et al. 2003).

AIZOACEAE (PREVIOUSLY MESEMBRYANTHEMACEAE)

Amphibolia hutchinsonii **(L. Bolus) H. E. K. Hartmann.**

Amphibolia hutchinsonii is native to the Cape Floristic Region of South Africa. Brown et al. (2003) reported that seed germination was increased by 1350% when its seeds were treated with aerosol smoke (Brown and Botha 2004).

Carpanthea pomeridiana **(L.) N. E. Br.**

GOLDEN CARPET

Golden carpet is a species that commonly occurs in fire-prone areas of South Africa's fynbos. Smoke water extracts produced by burning green fynbos plants and pumping the smoke through stream water did not significantly promote germination in this succulent (Pierce et al. 1995).

Caryotophora skiatophytoides **Leistn.**

Caryotophora skiatophytoides is a species common to the fire-prone areas of South Africa's fynbos. Smoke water extracts produced by burning green fynbos plants and funneling the smoke through stream water did not significantly promote germination in this species (Pierce et al. 1995).

Cleretum papulosum **(L. f.) L. Bolus**

Cleretum papulosum is a non-fire-prone succulent karoo species that occurs in South Africa and elsewhere in the world. Smoke water extracts produced by burning green fynbos plants and funneling the smoke through stream water did not significantly promote germination in this species (Pierce et al. 1995).

Conicosia pugioniformis **(L.) N. E. Br. ssp.** muirii **N. E. Br.**

PIG SALAD

Pig salad is common in the fire-prone areas of South Africa's fynbos. Smoke water extracts did not significantly promote germination in this species (Pierce et al. 1995).

Drosanthemum bellum **L. Bolus**

YELLOW-PINK BELL VYGIE

This succulent occurs in the Cape Floristic Region of South Africa. Aerosol smoke had no effect on its seed germination (Brown et al. 2003).

Drosanthemum speciosum **(Haw.) Schwantes**

DEW FLOWER ICE PLANT

Dew flower ice plant is common to the karoo ecosystems of South Africa. Pierce *et al.* (1995) were able to germinate 48% of the seeds by treating them with smoke water. Germination for the control was only 2%. Germination was also promoted when the seeds were treated with aerosol smoke (Brown et al. 2003; Brown and Botha 2004).

Drosanthemum stokoei **L. Bolus**

Drosanthemum stokoei occurs in the fire-prone areas of South Africa's fynbos. Smoke water extracts produced by burning green fynbos plants did not significantly promote germination in this species (Pierce et al. 1995).

Drosanthemum thudichumii **L. Bolus**

Brown *et al.* (2003) showed that the germination of this South African species significantly increased by 1,000% in response to aerosol smoke treatment (see also Brown and Botha 2004).

Erepsia anceps **(Haw.) Schwantes**

VYGIE

Erepsia anceps is native to the South African Cape Floristic Region. Pierce et al. (1995) observed that germination was promoted by smoke water. Germination, however, remained low, with only 10% germination in the treated seeds versus 1% germination for non-treated seed. Germination was also promoted when the seeds were treated with aerosol smoke (Brown et al. 2003; Brown and Botha 2004).

Erepsia aspera **(Haw.) L. Bolus**

Erepsia aspera occurs in the fire-prone areas of South Africa's fynbos. Smoke water extracts produced by burning green fynbos plants did not significantly promote germination in this species (Pierce et al. 1995).

Erepsia lacera **(Haw.) Liede**

PAARL ROOSVYGIE (AFR.)

Like *E. aspera*, paarl roosvygie occurs in the fire-prone areas of South Africa's fynbos. Smoke water extracts did not significantly promote germination in this species (Pierce et al. 1995).

Lampranthus aureus **(L.) N. E. Br.**

GOLDEN VYGIE

Golden vygie is native to the South African Cape Floristic Region. Germination was significantly promoted by smoke water (Pierce et al. 1995) or aerosol smoke (Brown and Botha 2004). Germination was, however, low (Pierce et al. 1995).

Lampranthus austricolus **L.**

This species is native to the South African Cape Floristic Region. Germination of its seed was not significantly promoted by 30 minutes exposure to aerosol smoke (Brown and Botha 2004).

Lampranthus bicolor **(L.) Jacobsen**

This species is native to the South African Cape Floristic Region. Germination of its seeds was not significantly promoted by 30 minutes exposure to aerosol smoke (Brown and Botha 2004).

Lampranthus haworthii **(Donn) N. E. Br.**

PURPLE VYGIE

This species is native to the South African Cape Floristic Region. Germination was significantly promoted by treating its seeds with aerosol smoke (Brown and Botha 2004).

Lampranthus multiradiatus **(Jacq.) N. E. Br.**

ROOSVYGIE (AFR.).

This species is native to the South African Cape Floristic Region. Germination was significantly promoted by treating its seeds with aerosol smoke (Brown and Botha 2004).

Lampranthus promontorii **N. E. Br.**

Like *L. austricolus* and *L. bicolor*, this species occurs in the fire-prone areas of South Africa's fynbos. Smoke water extracts produced by burning green fynbos plants and funneling the smoke through stream water did not significantly promote germination in this species (Pierce et al. 1995).

Leipoldtia schultzei **(Schltr. & Diels) Friedrich**

KUSSINGVYGIE (AFR.)

This succulent occurs in the Cape Floristic Region of South Africa. Aerosol smoke had no effect on its seed germination (Brown et al. 2003).

Oscularia deltoides **(L.) Schwantes**

DASSIEVYGIE (AFR.)

Dassievygie occurs in the fire-prone areas of South Africa's fynbos. Smoke water extracts did not significantly promote germination in this species (Pierce et al. 1995).

Ruschia caroli **(L. Bolus) Schwantes**

SHRUBBY DEWPLANT

Shrubby dewplant is native to the South African Cape Floristic Region. Seed responded positively to smoke water application. Germination of 70% was achieved, which may be due to the scarification prior to the smoke water application (Pierce et al. 1995). Physical dormancy may have prevented the smoke water from being as effective in promoting germination in other species. Germination of this species could also be promoted by exposing the seeds to aerosol smoke (Brown and Botha 2004).

Ruschia macrowanii **(L. Bolus) Schwantes**

BOSVYGIE (AFR.)

Bosvygie occurs in the fire-prone areas of South Africa's fynbos and elsewhere in the world. Smoke water extracts produced by burning green fynbos plants did not significantly promote germination in this species (Pierce et al. 1995).

Ruschia multiflora **(Haw.) Schwantes**

Ruschia multiflora is native to the South African Cape Floristic Region. Germination was significantly promoted by smoke water (Pierce et al. 1995) or aerosol smoke (Brown et al. 2003; see also Brown and Botha 2004).

Ruschia promontorii **L. Bolus**

Like *R. multiflora*, this species is native to the South African Cape Floristic Region. Seed germination was significantly increased following smoke water treatment (Pierce et al. 1995). This species is considered endangered and is in danger of extinction (IUCN 2013).

Ruschia sarmentosa **(Haw.) Schwantes**

Like *R. multiflora* and *R. promontorii*, this species is also a native of the Cape Floristic Region of South Africa. Seed germination was significantly improved when treated with a smoke water solution (Pierce et al. 1995).

Skiatophytum tripolium **(L.) L. Bolus**

PLATBAARVYGIE (AFR.)

Platbaarvygie occurs in the fire-prone areas of South Africa's fynbos. Smoke water did not significantly promote germination in this species (Pierce et al. 1995).

AMARANTHACEAE

Chenopodium album **L.**

LAMB'S QUARTERS

Lamb's quarters is native to Europe, and is a weed in Africa, Australasia and North America. Seed germination in this species was promoted when its seeds were soaked for 24 hours in smoke water (Daws et al. 2007).

Chenopodium carinatum **R. Br.**

GREEN GOOSEFOOT

The green goosefoot is native to the east coast of Australia, but occurs in other parts of world as an exotic plant. Soil sample were collected across forest edges between subtropical rainforests and eucalypt forests in the Lamington National Park of Queensland, Australia, and exposed for 60 minutes to cool aerosol smoke. There was no improvement in germination (Tang et al. 2003).

Einadia nutans **(R. Br.) A. J. Scott**

CLIMBING SALTBUSH

This saltbush occurs throughout Australia. Sixty minutes of aerosol smoke treatment of soil samples containing the seeds of this species, collected in the Lamington National Park of Queensland, Australia, had no effect on germination (Tang et al. 2003).

Ptilotus auriculifolius **(Moq.) F. Muell.**

EAR-LEAVED MULLA MULLA

Freshly collected seeds of this Australian native plant were exposed for 60 minutes to cool aerosol smoke, but germination was not significantly promoted (Roche et al. 1997b).

Ptilotus axillaris **(Benth.) F. Muell.**

Like *P. auriculifolius*, germination of the freshly collected seeds of this Australian plant, when exposed to cool aerosol smoke for 60 minutes, was not significantly promoted (Roche et al. 1997b).

Ptilotus drummondii **(Moq.) F. Muell.**

NARROWLEAF MULLA MULLA

Freshly collected seeds of this Australian native species were exposed to cool aerosol smoke for 60 minutes with no significant promotion in germination (Roche et al. 1997b).

Ptilotus exaltatus **Nees**

PINK MULLA MULLA

This species is widespread throughout the semiarid and arid zones of Australia. Freshly collected seeds were treated exposed to cool aerosol smoke for 60 minutes, some of which had received other treatments, but germination was not significantly promoted (Roche et al. 1997b).

Ptilotus helipteroides **F. Muell.**

HAIRY MULLA MULLA

This species is widespread throughout the semiarid and arid areas of Australia and commonly occurs in grasslands, herblands and mulga woodlands (Bean 2008). Freshly collected seeds did not significantly respond to cool aerosol smoke treatments of 60 minutes duration (Roche et al. 1997b).

Ptilotus manglesii **(Lindl.) F. Muell.**

POM POMS

Pom poms occur naturally in the jarrah (*Eucalyptus marginata*) forests of Western Australia. Neither aerosol smoke nor smoke water treatments significantly promoted germination of its seed (Norman et al. 2006).

Ptilotus nobilis **(Lindl.) F. Muell.**

YELLOWTAILS

This species occurs in several Australian east coast states. Freshly collected seeds were exposed for 60 minutes to cool aerosol smoke with no significant effect on germination (Roche et al. 1997b).

Ptilotus polystachys **F. Muell.**

Ptilotus polystachys is a native of the Western Australian eucalypt woodland, spinifex grassland and tropical savanna communities. Ex situ application of aerosol smoke to sown seeds had a significant inhibitory effect on final germination percentage (Roche et al. 1997b).

Ptilotus rotundifolius **(F. Muell) F. Muell.**

ROYAL MULLA MULLA

Freshly collected seeds of this Australian native species were exposed for 60 minutes to cool aerosol smoke with no significant effect on germination (Roche et al. 1997b).

ANACARDIACEAE

Litera caustica **(Mol.) H. & A.**

Thirty minutes exposure to cool aerosol smoke significantly inhibited germination in this woody species of the Mediterranean matorral of central Chile (Gómez-González et al. 2008).

Malosma laurina **(Nutt.) Nutt. ex Abrams**

LAUREL SUMAC

Laurel sumac occurs in the chaparral and western hardwood forests along the Pacific coast of California. Charate inhibited germination except when its seeds received a 5-minute heat shock treatment of 120°C, after which the charate promoted germination (Keeley 1987).

Rhus integrifolia **(Nutt.) Benth. & Hook. f. ex Brewer & S. Watson**

LEMONADE SUMAC

Lemonade sumac commonly occurs on the mesas and hillslopes of southern California, USA. Under lighted conditions, charate inhibited germination, except when the seeds received a 5-minute heat shock treatment of 120°C. Germination was then promoted. Under dark conditions, germination was promoted by charate treatment regardless of whether or not the seeds had been exposed to heat (Keeley 1987).

Rhus schmidelioides **Schltdl**

RHUS

A variety of different fire cues were tested on the seeds of this species, which were collected from a mixed forest located in a mountainous subtropical area of Mexico (Zuloaga-Aguilar et al. 2011). These included heat shock (100:15 sand: water (w:w) substrate at 120°C for 5 minutes), soaking the seeds in smoke water for 3 hours (prepared by burning 150 g of mixed forest litter and bubbling the resultant smoke into 1.5 L of distilled water, and adjusting the pH to 5 using sodium hydroxide) or ash (1.5 g of fine ash was added to the agar plates used to germinate the seeds). Combinations of these treatments were also tested for their effects on seed germination. The smoke water treatment inhibited germination of this species, while heat shock increased it. None of the other treatments had any effect.

Rhus tomentosa **L.**

This species occurs in the Cape Floristic Region of South Africa. An application of aerosol smoke did not improve seed germination (Brown et al. 2003).

Rhus trichocarpa **Miq.**

JAPANESE SUMAC

The effects of aerosol smoke, heat, darkness, cold stratification, and combinations of smoke with each of the three other treatments on seed germination were examined in this study (Tsuyuzaki and Miyoshi 2009). Smoke was produced by burning Timothy hay (*Phleum pratense*), which was pumped through a 3.5 m cooling tube into a smoke chamber for approximately 5 minutes. The seeds were exposed to the smoke for 60 minutes. Those seeds exposed also to heat were incubated at 75°C for 25 minutes. The cold stratification process took 1 month, during which the seeds remained in an incubator set to 4°C. Where the dark treatment was concerned, the seeds were maintained in total darkness for the entire germination period. None of the treatments tested had any effect on germination, which was low even for the control group (3%).

Rhus trilobata **Nutt.**

SKUNKBUSH SUMAC

This sumac has a wide distribution across the western parts of the USA and Canada. Charate promoted germination of its seeds under light and dark conditions. The best germination results were, however, achieved under light conditions (Keeley 1987).

Toxicodendron diversilobum **(Torr. & Gray) Greene**

PACIFIC POISON OAK

Pacific poison oak occurs in California, Nevada, Oregon, and Washington. Like *R. trilobata* above, treatment with charate promoted germination under light and dark conditions, with the best results being achieved under light conditions (Keeley 1987).

APIACEAE

Actinotus helianthi **Labill.**

FLANNEL FLOWER

Flannel flower, which is endemic to Australia, occurs from central Queensland down to the south coast and western slopes of New South Wales. *Actinotus*

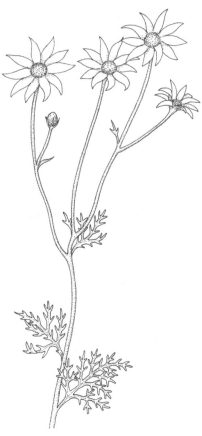

Actinotus helianthi

helianthi grows commonly in coastal heaths, scrub, and dry sclerophyll forests. Germination was significantly enhanced by aerosol smoke treatments according to an ex situ study by Roche et al. (1997b).

Actinotus leucocephalus **Benth.**

FLANNEL FLOWER

Flannel flower is common in the proteaceous heathlands, banksia woodlands, and jarrah forests of Western Australia. Significant increases in germination (30%) occurred when its seeds were stored at 50°C (10–90 days) prior to being sown and then treated for 60 minutes with aerosol smoke (Tieu et al. 1999, Tieu et al. 2001b). Baker et al. (2005) reported that germination also increased significantly when the seeds were treated with smoke water. Those seeds did require, however, pretreatment, such as manual scarification and heating (70°C for 60 minutes), prior to incubation in smoke water. Interactions between heat and smoke and smoke and exposure time also resulted in improved germination for this species, but extremely high temperatures were not significantly better (Tieu et al. 2001a). In contrast, Roche et al. (1997b) reported that treating freshly collected seeds for 60 minutes with cool aerosol smoke did not significantly improve germination.

Alepidea natalensis **J. M. Wood & M. S. Evans.**

Treatment with smoke water significantly improved seed germination and seedling vigor in this highly valued medicinal plant (Mulaudzi et al. 2009).

Angelica sylvestris **L.**

WILD ANGELICA

This European herb is a cool temperate arable weed that occurs in parts of Australia. A commercially available smoke water solution (Seed Starter®) did not significantly improve seed germination (Adkins and Peters 2001).

Daucus glochidiatus **(Labill.) Fisch., C. A. Mey. & Avé-Lall.**

AUSTRALIAN CARROT

Aerosol smoke treatments of broadcast seeds in rehabilitated mine sites in the southwest of Western Australia had no effect on their germination (Roche et al. 1997a). Soil samples containing the seeds of this species, when collected across forest edges between subtropical rainforest and eucalypt forest in the Lamington National Park of Queensland, Australia, were similarly exposed to cool aerosol smoke for 60 minutes and also resulted in no significant improvement to germination (Tang et al. 2003).

Daucus pusillus **Michx.**

AMERICAN WILD CARROT

This species is widely distributed across the USA, western Canada, northern Mexico, Brazil, Argentina and Chili. Keeley and Keeley (1987) showed that charate (0.5 g on filter paper) inhibited the germination of this annual.

Heracleum sphondylium **L.**

HOGWEED

Hogweed is a cool temperate arable weed that occurs in parts of Australia. A commercially available smoke water solution "Seed Starter" did not significantly improve seed germination in this species (Adkins and Peters 2001).

Hydrocotyle callicarpa **Bunge**

SMALL PENNYWORT

Hydrocotyle callicarpa is widely distributed throughout southern Australia. In a study of the effects of aerosol smoke on in situ seed germination, the mean density of germinants per plot increased from 0.31 in the control to 20.34 in aerosol smoke treated plots (Roche et al. 1997a). Smoke water had no effect, however. Enright and Kintrup (2001), in contrast, observed increased numbers of small pennywort germinants in response to smoke water treatments of the soil seed bank.

Hydrocotyle pedicellosa **F. Muell.**

LARGE PENNYWORT

Soil samples containing seeds of this species, collected across forest edges between subtropical rainforest and eucalypt forest in the Lamington National Park of Queensland, Australia, were exposed to cool aerosol smoke for 60 minutes with no significant improvement to germination (Tang et al. 2003).

Hydrocotyle **sp.**

Soil samples from the Eden Burning Study Area, a dry sclerophyll forest in the Yalumba State Forest of New South Wales, Australia, were collected and air dried to test the effects of heat, smoke, and an interaction between the two cues on seeds from the seed bank. Samples exposed to heat treatment were incubated at 80°C for 60 minutes while those exposed to smoke were incubated in a room, where smoke was generated for 120 minutes. None of the treatments had any effect on the germination of the seeds of this species (Penman et al. 2008).

Lomatium dasycarpum **(Torr. & Gray)**

WOOLLY FRUIT DESERT PARSLEY

Germination of this parsley, native to the chaparral of California, USA, and northern Mexico, was inhibited by a charred wood treatment (Keeley and Keeley 1987).

Mulinum spinosum **Pers.**

HIERBA NEGRA

The germination of hierba negra seeds, when treated with a combination of heat (80°C for 5 minutes) and aerosol smoke for 60 minutes, was not significantly affected (Gonzalez and Ghermandi 2012). This species is native to the grasslands of northwestern Patagonian.

Pentapeltis peltigera **(Hook.) Bunge**

A 1 L/m² application of undiluted smoke water to the soil surface of a rehabilitated bauxite mine in the southwest of Western Australia had no significant effect on the germination of seeds from the soil seed bank (Roche et al. 1997a). Sixty minutes of cool aerosol smoke also had no effect on germination (Roche et al. 1997b).

Platysace compressa **(Labill.) C. Norman**

TAPEWORM PLANT

Tapeworm plant occurs in the coastal limestone and sand dunes of the proteaceous heathland communities, granite hills and rocky outcrops of banksia woodland and jarrah forest communities. Roche et al. (1997a) reported that germination of broadcast seeds was not significantly improved with aerosol smoke or smoke water.

Platysace lanceolata **(Labill.) Druce**

SHRUBBY PLATYSACE

Soil samples from the Eden Burning Study Area, a dry sclerophyll forest in the Yalumba State Forest of New South Wales, Australia, were collected and air dried to test the effects of heat, smoke, and an interaction between the two cues on seeds from the seed bank. Samples exposed to heat treatment were incubated at 80°C for 60 minutes while those exposed to smoke were incubated in a room, where smoke was generated for 120 minutes. Both heat and smoke on their own significantly improved germination of the seeds of this species while an interaction between the two cues had no effect (Penman et al. 2008).

Platysace tenuissima **(Benth.) C. Norman**

Aerosol smoke treatment of broadcast seed, as well as a 1 L/m² application of undiluted smoke water to the soil surface of a rehabilitated bauxite mine in the southwest of Western Australia had no significant effect on the germination on the seeds of this species (Roche et al. 1997a).

Trachymene pilosa **Sm.**

NATIVE PARSNIP

Native parsnip occurs throughout the Southwest Botanical Province of Western Australia. Smoke water applications to the soil seed bank significantly promoted germination of this species on a bauxite mine site in a jarrah forest that had been undergoing rehabilitation (Roche et al. 1997a). Aerosol smoke treatments of broadcast seed, in contrast, did not significantly improve germination.

Xanthosia candida **(Benth.) Steud.**

Xanthosia candida is common to the proteaceous heathlands, banksia woodlands, and jarrah forests of southwestern Western Australia. The germination of this species was significantly promoted when in situ aerosol smoke was applied to the soil surface for a period of 60 minutes (Roche et al. 1997a). Smoke water had no effect.

Xanthosia huegelii **(Benth.) Steud.**

HEATH XANTHOSIA

Smoke water application to the soil seedbank of this *Xanthosia* increased germination significantly from a mean of 4.30 seedlings in the control plots to 11.17 seedlings in the treated plots of areas of a bauxite mine site undergoing rehabilitation (Roche et al. 1997a). Also, an ex situ study revealed improved germination in response to aerosol smoke treatment of sown seeds (Roche et al. 1997b). Dixon et al. (1995) previously reported that 90 minutes of cold smoke treatment had no effect on seed germination. *Xanthosia heugelii* is a native of the proteaceous heathlands, banksia woodlands, and jarrah forests of southwestern Western Australia.

Xanthosia **sp.**

Foliar smoke water sprays, with concentrations of 50 and 100 mL/m², did not significantly promote germination of seeds of an unidentified species of *Xanthosia* occurring in an intact banksia woodland 20 km south of Perth, Western Australia (Lloyd et al. 2000).

Xanthosia tridentata **DC.**

HILL XANTHOSIA

Soil samples from the Eden Burning Study Area, a dry sclerophyll forest in the Yalumba State Forest of New South Wales, Australia, were collected and air dried to test the effects of heat, smoke, and an interaction between the two cues on seeds from the seed bank. Samples exposed to heat treatment were incubated at 80°C for 60 minutes while those exposed to smoke were incubated in a room, where smoke was generated for 120 minutes. Only the smoke treatment significantly improved the germination of the seeds of this species (Penman et al. 2008). Neither heat nor the interaction between the two cues had any effect.

APOCYNACEAE

Alyxia buxifolia **R. Br.**

Alyxia buxifolia has a wide distribution across southern Australia. Germination was significantly greater when *A. buxifolia* seeds were treated with aerosol smoke, but only after the seeds had been stored in soil for 12 months. Germination responses of 0%, 2.6%, and 34.2% occurred after no treatment, smoke treatment, and soil storage followed by smoke treatment, respectively (Roche et al. 1997b).

Marsdenia **sp.**

Soil samples containing seeds of an unidentified species of *Marsdenia*, collected across forest edges between subtropical rainforest and eucalypt forest in the Lamington National Park of Queensland, Australia, were exposed for 60 minutes to cool aerosol smoke with no significant improvement to germination (Tang et al. 2003).

ASCLEPIADACEAE

Asclepias tuberosa **L.**

BUTTERFLY MILKWEED

The seeds of this species, collected from a tallgrass prairie in the midwestern United States (Jefferson et al. 2007) and a ponderosa pine (*Pinus ponderosa*) forest in northern Arizona, United States (Abella and Springer 2009), did not significantly respond to aerosol smoke (Jefferson et al. 2007) or concentrated aqueous smoke treatment (Wright's Brand, Roseland, New Jersey, United States) (Abella and Springer 2009).

ASTERACEAE

Acourtia microcephala **DC.**

Acourtia microcephala is native to the Californian chaparral of the United States. Seed germination was inhibited by an application of charred wood extract (Keeley and Keeley 1987).

Ageratina adenophora **(Spreng) R. M. King & H. Rob.**

CROFTON WEED

Soil samples containing the seeds of this native of Mexico, were collected across forest edges between subtropical rainforest and eucalypt forest in the Lamington National Park of Queensland, Australia, and exposed for 60 minutes to cool aerosol smoke with no significant increase in germination (Tang et al. 2003).

Agoseris heterophylla **(Nutt.) Greene**

ANNUAL AGOSERIS

Agoseris heterophylla commonly occurs west of the Rocky Mountains of the United States. Charred wood (0.5 g on filter paper) significantly increased germination from 62% in the control (no treatment) to 92% when seeds were treated (Keeley and Keeley 1987).

Ammobium alatum **R. Br.**

WINGED EVERLASTING

Ammobium alatum is common to the grassy woodlands of the New England Tableland of New South Wales, Australia. Smoke water treatments had no effect on the germination of this species (Clarke et al. 2000).

Anaphalis margaritacea **(L.) Benth. & Hook. f. var.** angustior **(Miq.) Nakai**

PEARLY EVERLASTING

The effects of aerosol smoke, heat, darkness, cold stratification, and combinations of smoke with each of the three other treatments on seed germination were examined in this study (Tsuyuzaki and Miyoshi 2009). Smoke was produced by burning Timothy hay (*Phleum pratense*), which was pumped through a 3.5 m cooling tube into a smoke chamber for approximately 5 minutes. The seeds were exposed to the smoke for 60 minutes. Those seeds exposed also to heat were incubated at 75°C for 25 minutes. The cold stratification process took 1 month, during which the seeds remained in an incubator set to 4°C. Where the dark treatment was concerned, the seeds were maintained in total darkness for the entire germination period. Compared to the control group, in which 94% germination was achieved, three treatments significantly inhibited the seeds of this species: smoke (43% germination), dark (36% germination), and smoke combined with cold stratification (74% germination). The seeds of this species display physiological dormancy.

Angianthus tomentosus **J. C. Wendl.**

HAIRY CUP-FLOWER

Angianthus tomentosus commonly grows in the arid regions of Western Australia, South Australia, Victoria and New South Wales in Australia. Karrikinolide (100 ppb) and smoke water (1:10 dilution; prepared in Dixon et al. 1995) promoted the germination of this species (Merritt et al. 2006). Gibberellic acid (1,000 ppm) was equally as effective as karrikinolide (1, 10, or 100 ppb) at promoting germination under light and dark conditions and at temperature regimes of 26/13°C and 33/18°C (Merritt et al. 2006). However, at lower temperatures, gibberellic acid was less effective in promoting germination than karrikinolide.

Antennaria rosulata **Rydb.**

KAIBAB PUSSYTOES

The seeds of this species, collected from a ponderosa pine (*Pinus ponderosa*) forest in northern Arizona, United States, did not significantly respond to concentrated aqueous smoke treatment (Wright's Brand, Roseland, New Jersey) (Abella and Springer 2009).

Arctotheca calendula **(L.) Levyns**

CAPEWEED

Capeweed is native to the Cape Province of South Africa and has naturalized in Mediterranean regions of Europe, Australia, New Zealand, the United States, and the Azores. Treatment of seeds with karrikinolide (0.67 μM) promoted germination in this species (Stevens et al. 2007). Germination remained low, however (approximately 10%).

Arctotis acaulis **L.**

AFRICAN DAISY

This species occurs in the Cape Floristic Region of South Africa. An application of aerosol smoke did not significantly improve seed germination (Brown et al. 2003).

Arctotis stoechadifolia **(P. J. Bergius)**

BLUE-EYED AFRICAN DAISY

Thirty minutes of aerosol smoke treatment significantly promoted germination in this species (Brown et al. 2003; Brown and Botha 2004), which is native to the Cape Floristic Region of South Africa.

Arnica chamissonis **Less.**

The seeds of this species, collected from a ponderosa pine (*Pinus ponderosa*) forest in northern Arizona, United States, did not significantly respond to concentrated aqueous smoke treatment (Wright's Brand, Roseland, New Jersey) (Abella and Springer 2009).

Arnoglossum atripilicifolium **(L.) H. E. Rob.**

PALE INDIAN PLANTAIN

The seeds of this species, collected from a tallgrass prairie in the Midwest of the United States, did not germinate in response to aerosol smoke treatments of 1, 10, or 60 minutes duration (Jefferson et al. 2007).

Artemisia californica **Less.**

CALIFORNIA SAGEBRUSH

Endemic to California, this sagebrush occurs in the chaparral, western hardwoods, pinyon-juniper, and annual grasslands ecosystems. Under light conditions, powdered charred wood (charate) treatment had no effect on germination. Under dark conditions, germination increased as a result of charate treatment. Germination was greater under light conditions without charate treatment germination than under dark conditions with charate treatment (Keeley 1987).

Artemisia ludoviciana **Nutt.**

WHITE SAGEBRUSH

The emergence of seedlings was significantly increased after the seeds of this species, collected from a frequently burned ponderosa pine forest (*Pinus ponderosa*) in northern Arizona, were treated with concentrated aqueous smoke (Wright's Brand, Roseland, New Jersey) (Abella and Springer 2009).

Artemisia montana **(Nakai) Pamp.**

MOUNTAIN MUGWORT

The effects of aerosol smoke, heat, darkness, cold stratification, and combinations of smoke with each of the three other treatments on seed germination were examined in this study (Tsuyuzaki and Miyoshi 2009). Smoke was produced by burning Timothy hay (*Phleum pratense*), which was pumped through a 3.5 m cooling tube into a smoke chamber for approximately 5 minutes. The seeds were exposed to the smoke for 60 minutes. Those seeds exposed also to heat were incubated at 75°C for 25 minutes. The cold stratification process took 1 month, during which the seeds remained in an incubator set to 4°C. Where the dark treatment was concerned, the seeds were maintained in total darkness for the entire germination period. The dark treatment alone reduced germination from 90% to 6%. Significant inhibition also occurred when the smoke dark treatments were combined, decreasing to 43%. The seeds of this species, which occurs in China and Japan, display physiological dormancy.

Aster ageratoides Turcz. **ssp.** ovatus **(Franch. & Sav.) Kitam. var.** yezonensis

NO KO NGIKU (JAPANESE)

The effects of aerosol smoke, heat, darkness, cold stratification, and combinations of smoke with each of the three other treatments on seed germination were examined in this study (Tsuyuzaki and Miyoshi 2009; see *Artemisia montana* above for details of the tests performed). The dark treatment significantly reduced germination for this species, with 7% germination compared to the control (28%). None of the other treatments had any significant effect on germination.

Baccharis vernalis **F. H. Hellw.**

BACCHARIS

Thirty minutes exposure to cool aerosol smoke significantly promoted germination in this woody species of the Mediterranean matorral of central Chile (Gómez-González *et al.* 2008).

Bahia dissecta **Britton**

RAGLEAF BAHIA

The seeds of this species, collected from a ponderosa pine (*Pinus ponderosa*) forest in northern Arizona did not significantly respond to concentrated aqueous smoke treatment (Wright's Brand, Roseland, New Jersey) (Abella and Springer 2009).

Boltonia decurrens **(Torr. & Gray) Alph. Wood.**

CLASPINGLEAF DOLL'S DAISY

Seed of this species, collected from a tallgrass prairie in the Midwest of the United States, did not germinate in response to aerosol smoke treatments of 1, 10 or 60 minutes duration (Jefferson et al. 2007).

Bracteantha bracteata **(Vent.) Anderb. & Haegi**

PAPER DAISY

Paper daisy is common to the grassy woodlands of the New England Tableland of New South Wales, Australia, and elsewhere in eastern Australia. Smoke water treatments had no effect on the germination of this species (Clarke et al. 2000).

Cassinia longifolia **R. Br.**

SHINY CASSINIA

Soil samples from the Eden Burning Study Area, a dry sclerophyll forest in the Yalumba State Forest of New South Wales, Australia, were collected and air dried to test the effects of heat, smoke, and an interaction between the two cues on seeds from the seed bank. Samples exposed to heat treatment were incubated at 80°C for 60 minutes while those exposed to smoke were incubated in a room, where smoke was generated for 120 minutes. The heat treatment on its own promoted the germination of the seeds of this species, but was only marginally significant (Penman et al. 2008). Neither the smoke treatment nor interaction between the two cues had any effect.

Cassinia quinquefaria **R. Br.**

LONG-LEAVED CASSINIA

Cassinia quinquefaria, like *B. bracteata* above, is also common to the grassy woodlands of the New England Tableland of New South Wales, Australia. Smoke water treatments had no effect on the germination of this species either (Clarke et al. 2000).

Chaenactis artemisiaefolia **(Harv. & Gray ex Gray) Gray**

WHITE PINCUSHION

White pincushion is native to California's chaparral communities. Keeley et al. (1985) first discovered that powdered charred wood promoted germination on this species. Keeley and Fotheringham (1998a) later demonstrated that aerosol smoke treatments (5 or 8 minutes exposure) also improved its germination. The global threat status of this species is rare or considered at risk; the local threat status is indeterminate (IUCN 2013).

Chaenactis glabriuscula **DC.**

YELLOW PINCUSHION

Yellow pincushion is native to the coastal sage scrub, oak woodland, and chaparral ecosystems of California and northern Mexico. Fotheringham et al. (1995) reported that smoke stimulated germination in this species. The global threat status of this species is rare or considered at risk, while the local threat status remains indeterminate (IUCN 2013).

Chromolaena odorata **(L.) R. M. King & H. Rob.**

SIAM WEED

This species is native to North America, and is considered a weed in northern Queensland, Australia. The seed germination of Siam weed, from primary forests near Townsville and secondary forests of the South Johnstone River Catchment in northern Queensland, was significantly increased when exposed for 10 minutes to aerosol smoke (Djietror et al. 2011).

Chrysanthemum segetum **(L.) Fourr.**

CORNDAISY

This European herb is a common weed in disturbed areas. Germination was promoted when its seeds were exposed for 24 hours to smoke water (Capeseed, Cape Town, South Africa) or 10^{-7} M of karrikinolide (Daws et al. 2007).

Chrysocephalum baxteri **(A. Cunn ex DC.) Anderb.**

FRINGED EVERLASTING

Soil samples from the Eden Burning Study Area, a dry sclerophyll forest in the Yalumba State Forest of New South Wales, Australia, were collected and air dried to test the effects of heat, smoke, and an interaction between the two cues on seeds from the seed bank. Samples exposed to heat treatment were incubated at 80°C for 60 minutes while those exposed to smoke were incubated in a room, where smoke was generated for 120 minutes. Germination was significantly increased following smoke treatment, but was inhibited by heat (Penman et al. 2008). An increase in germination due to an interaction between heat and smoke was only marginally significant.

Chyrsocoma coma-**aurea L.**

This species occurs in the Cape Floristic Region of South Africa. An application of aerosol smoke did not significantly improve seed germination (Brown et al. 2003).

Cirsium vulgare **(Savi) Ten.**

BULL THISTLE

Soil samples, containing seeds of this native of Europe, Africa, and Asia, were collected across forest edges between subtropical rainforest and eucalypt forest in the Lamington National Park of Queensland, Australia, and exposed to cool aerosol smoke for 60 minutes. There was no improvement in germination (Tang et al. 2003).

Conyza canadensis **(L.) Cronquist**

CANADIAN HORSEWEED

Soil samples containing the seeds of this North and Central American horseweed, which were collected across forest edges between subtropical rainforest and eucalypt forest in the Lamington National Park of Queensland, Australia, were exposed to cool aerosol smoke for a period of 60 minutes with no significant improvement to germination (Tang et al. 2003). This species was formerly known as *Erigeron canadensis*.

Coreopsis basalis **(A. Dietr.) S. F. Blake**

GOLDEN MAN COREOPSIS

Following an aerosol smoke treatment of 8 minutes, heat treatment of either 30 or 60 seconds in an oven at 100°C and wet, cold stratification for 1 month at 4°C, there was a significant decrease in germination (Schwilk and Zavala 2012). A similar response was reported for seeds stratified in dry, cold environment at 4°C for 1 month with a relative humidity of 10%. If there was no stratification, germination significantly increased following the smoke treatment (from 41% to 78%).

Coreopsis lanceolata **L.**

LANCELEAF TICKSEED

Lanceleaf tickseed is native to the central and eastern states of the United States and eastern Canada. This species has become a weed in many parts of the world. Jefferson et al. (2007) reported that germination of its seeds doubled when fumigated with aerosol smoke for 10 minutes. Schwilk and Zavala (2012) also tested the effects of aerosol smoke on this species. They exposed the seeds to either 4 or 8 minutes of smoke, incubated them in an oven at 100°C for for either 30 or 60 seconds and then subjected them to wet, cold stratification for 1 month at 4°C. The results of this study revealed there was no significant effect on germination following 4 minutes of smoke treatment and a significant decrease in it after 8 minutes exposure to smoke regardless of temperature. When the seeds were stratified in a dry, cold environment at 4°C for 1 month, with a relative humidity of 10%, there was no significant effect on germination due to the smoke treatment (Schwilk and Zavala 2012).

Coreopsis tinctoria **Nutt.**

PLAINS COREOPSIS

Following aerosol smoke treatments of 4 or 8 minutes, heat treatment of either 30 or 60 seconds in an oven at 100°C and wet, cold stratification for 1 month at 4°C, there was no germination in treatments or controls for this species (Schwilk and Zavala 2012). Even when the seeds were stratified in a dry, cold environment at 4°C for 1 month with a relative humidity of 10%, there was no significant effect on germination due to smoke treatment (Schwilk and Zavala 2012). Chou et al. (2012) tested smoke water, heat and combinations of both heat and smoke on the germination percentage and mean germination times of this species. Smoke treatments were applied by soaking seeds for 20 hours in the commercially available Regen 2000® smoke water solution, at concentrations of 1:5, 1:10, or 1:100 (v/v). Seeds were exposed to heat treatments of 50 or 80°C for 5 minutes. *Coreopsis tinctoria* germination percentages and mean germination times were significantly inhibited by 1:5, or 1:10 and 1:5 concentrations of smoke water, respectively. The seeds of this species did not respond to heat treatments alone, nor was there any significant interaction between heat and smoke treatments.

Corymbium glabrum **L. var.** glabrum

This species is native to the Cape Floristic Region of South Africa. Exposure of its seeds to aerosol smoke for 30 minutes promoted germination (Brown et al. 2003).

Corymbium laxum **Compton ssp.** bolusii **F. M. Weitz**

This species is also native to the Cape Floristic Region of South Africa. Seed germination was significantly promoted when its seeds exposed to aerosol smoke for 30 minutes (Brown et al. 2003).

Cotula turbinata **L.**

BRASS BUTTONS

Brass buttons occurs in the Cape Floristic Region of South Africa and elsewhere in the world, including Western Australia, where it is regarded an alien species. An application of aerosol smoke did not improve seed germination (Brown et al. 2003).

Crassocephalum crepidioides **(Benth.) S. Moore**

THICKHEAD

Soil samples containing the seeds of this African and Madagascan species, were collected across forest edges between subtropical rainforest and eucalypt forest in the Lamington National Park of Queensland, Australia, and exposed to cool aerosol smoke for a period of 60 minutes, to which there was no significant promotion in germination (Tang et al. 2003).

Cyanthillium cinereum **(L.) H. Rob.**

LITTLE IRONWEED

Little ironweed is common to southeast Asia, but also occurs in parts of the United States and Australia. Soil samples containing the seeds of this species were collected across forest edges between subtropical rainforest and eucalypt forest in the Lamington National Park of Queensland, Australia, and exposed to cool aerosol smoke for a period of 60 minutes with no significant improvement to germination (Tang et al. 2003).

Dimorphotheca nudicaulis **DC.**

OX-EYE DAISY

Germination of ox-eye daisy, a native to South Africa, increased 225% in response to 30 minutes of aerosol smoke treatment (Brown and Botha 2004). A previous and similar study revealed no significant germination following smoke treatment (Brown et al. 2003).

Disparago ericoides **(Berg.) Gaertn.**

Seed germination in this fynbos species of the Cape Floristic Region of South Africa was not significantly improved following 30 minutes of aerosol smoke treatment (Brown and Botha 2004).

Dittrichia viscosa **(L.) Greuter**

FALSE YELLOWHEAD

False yellowhead is native to southern Europe, northern Africa, and western Asia and has since become a weed in many parts of the world. Germination of *D. viscosa* seeds was promoted when they were exposed to aerosol smoke (10 or 20 minutes) (Pérez-Fernández and Rodríguez-Echeverría 2003).

Echinacea angustifolia **DC.**

BLACKSAMSON ECHINACEA

This echinacea is native to the prairies of the northwestern and central United States and western Canada. Germination of this species was significantly promoted by exposing the seeds to aerosol smoke for 1, 10, and 60 minutes (Jefferson et al. 2007). Germination percentages were approximately 70%, 80%, and 90%, respectively, in comparison to 55% for the control treatment. Schwilk and Zavala (2012) also tested the effects of aerosol smoke on the seeds of this species. They reported no significant improvement to germination once the seeds were exposed to smoke for 4 or 8 minutes, heat treatment at 30 or 60 seconds in an oven at 100°C, and following wet, cold stratification for 1 month at 4°C. Seed stratified in a dry, cold environment at 4°C for 1 month, and with a relative humidity of 10%, did not germinate either.

Echinacea atrorubens **Nutt.**

TOPEKA PURPLE CONEFLOWER

The seeds of Topeka purple coneflower, collected from a tallgrass prairie in the Midwest of the United States, did not germinate in response to aerosol smoke treatments of 1, 10, or 60 minutes duration (Jefferson et al. 2007).

Echinacea pallida **(Nutt.) Nutt.**

PALE PURPLE CONEFLOWER

Germination of *E. pallida*, native to the eastern and central United States, doubled in response after 60 minutes of exposure to aerosol smoke (Jefferson et al. 2007).

Echinacea paradoxa **(Norton) Britton**

BUSH'S PURPLE CONEFLOWER

Echinacea paradoxa has a limited distribution in Missouri and Arkansas in the United States, and has become a weed in many parts of the world. Jefferson et al. (2007) reported that germination of its seeds was promoted when they were fumigated with aerosol smoke for a period of 60 minutes. Fifty percent of the treated seed germinated in comparison to 25% for the control group.

Echinacea purpurea **(L.) Moench**

EASTERN PURPLE CONEFLOWER

Eastern purple coneflower commonly occurs throughout much of the eastern and central United States. Germination of this species was promoted by aerosol smoke (Jefferson *et al.* 2007). Ninety-five percent of seed germinated when they were exposed to aerosol smoke for 10 minutes in comparison to 80% for the control group. Schwilk and Zavala (2012) reported, in contrast, that the germination of this species' seeds, when exposed to an aerosol smoke treatment of 8 minutes, followed by heat treatment at 30 or 60 seconds in an oven at 100°C and wet, cold stratification for 1 month at 4°C, was inhibited. There was no significant effect on germination due to smoke treatment if the seeds were stratified in a dry, cold environment at 4°C for 1 month at a relative humidity of 10%.

Echinacea tennesseensis **(Beadle) Small**

TENNESSEE PURPLE CONEFLOWER

This endangered (IUCN 2013) coneflower occurs in Tennessee, USA. Jefferson et al. (2007) reported that approximately 90% of its seeds germinated when exposed to aerosol smoke for a period of 60 minutes, in comparison to 50% for the controls (no treatment).

Edmondia fasciculata **(Andrews) Hilliard**

Seed germination in this fynbos species of the Cape Floristic Region of South Africa was not significantly improved following 30 minutes of aerosol smoke treatment (Brown and Botha 2004).

Edmondia sesamoides **(L.) Hilliard**

EVERLASTING

Edmondia sesamoides is native to the Cape Floristic Region of South Africa. Aerosol smoke (30 minutes) significantly promoted germination (Brown et al. 2003; Brown and Botha 2004; Brown et al. 1995).

Emilia sonchifolia **(L.) DC.**

RED TASSEL FLOWER

This species occurs in eastern and southern Asia and western Pacific regions of the world. Soil samples containing seeds of this species were collected across forest edges between subtropical rainforest and eucalypt forest in the Lamington National Park of Queensland, Australia, and were exposed to cool aerosol smoke for 60 minutes no significant improvement in germination (Tang et al. 2003).

Erigeron speciosus **DC.**

ASPEN FLEABANE

The seeds of this species were collected from a frequently burned ponderosa pine forest (*Pinus ponderosa*) in northern Arizona, and were exposed to concentrated aqueous smoke (Wright's Brand, Roseland, New Jersey). This significantly increased the emergence of its seedlings (Abella and Springer 2009).

Eriocephalus africanus **L.**

WILD ROSEMARY

Eriocephalus africanus is native to the Cape Floristic Region of South Africa. Aerosol smoke (30 minutes) significantly promoted germination (Brown et al. 2003).

Eriophyllum confertiflorum **(DC.) A. Gray**

GOLDEN YARROW

Germination of this native to the chaparral of California and Arizona, USA, was promoted by powdered charred wood seed treatments (Keeley et al. 1985).

Euchiton gymnocephalus **(DC.) Anderb.**

CREEPING CUDWEED

Soil samples from the Eden Burning Study Area, a dry sclerophyll forest in the Yalumba State Forest of New South Wales, Australia, were collected and air dried to test the effects of heat, smoke, and an interaction between the two cues on seeds from the seed bank. Samples exposed to heat treatment were incubated at 80°C for 60 minutes while those exposed to smoke were incubated in a room, where smoke was generated for 120 minutes. Neither heat nor smoke had any effect on the germination of the seeds of this species, but an interaction between the two cues significantly inhibited it (Penman et al. 2008).

Euchiton involucratus **(G. Forst.) Holub**

STAR CUDWEED

Soil samples containing the seeds of this species, collected across forest edges between subtropical rainforest and eucalypt forest in the Lamington National Park of Queensland, Australia, were exposed for 60 minutes to cool aerosol smoke. There was no improvement in germination (Tang et al. 2003).

Euchiton sphaericus **(Willd.) Anderb.**

ANNUAL CUDWEED

Soil samples from the Eden Burning Study Area, a dry sclerophyll forest in the Yalumba State Forest of New South Wales, Australia, were collected and air dried to test the effects of heat, smoke, and an interaction between the two cues on seeds from the seed bank. Samples exposed to heat treatment were incubated at 80°C for 60 minutes while those exposed to smoke were incubated in a room, where smoke was generated for 120 minutes. None of the treatments had any effect on the germination of the seeds of this species (Penman et al. 2008).

Eupatorium chinense **L. ssp.** sachalinense **(F. Schm.) Kitam. ex Murata**

EUPATORIUM

The effects of aerosol smoke, heat, darkness, cold stratification, and combinations of smoke with each of the three other treatments on seed germination were examined in this study (Tsuyuzaki and Miyoshi 2009). Smoke was produced by burning Timothy hay (*Phleum pratense*), which was pumped through a 3.5 m cooling tube into a smoke chamber for approximately 5 minutes. The seeds were exposed to the smoke for 60 minutes. Those seeds exposed also to heat were incubated at 75°C for 25 minutes. The cold stratification process took 1 month, during which the seeds remained in an incubator set to 4°C. Where the dark treatment was concerned, the seeds were maintained in total darkness for the entire germination period. Germination percentage was recorded for each treatment. Germination of the control group was high, with 88% of the seeds germinating. However, when treated with smoke, almost no germination occurred (0%–0.5%). Seeds incubated in the dark also experienced significantly decreased germination (8%).

Euryops abrotanifolius **(L.) DC.**

MOUNTAIN RESIN BUSH

Seed germination in this fynbos species of the Cape Floristic Region of South Africa was not significantly improved following 30 minutes of aerosol smoke treatment (Brown and Botha 2004).

Euryops linearis **Harv.**

Euryops linearis is native to the Cape Floristic Region of South Africa. Aerosol smoke (30 minutes) significantly promoted germination (Brown et al. 2003; Brown and Botha 2004). This species is considered both rare and at risk at the local and global scale (IUCN 2013).

Euryops speciosissimus **DC.**

CLANWILLIAM EURYOPS

Like *E. linearis*, germination in this native fynbos species of the Cape Floristic Region of South Africa was significantly improved by 30 minutes of aerosol smoke treatment (Brown et al. 2003; Brown and Botha 2004). Brown and Botha (2004) also reported that some samples of this species did not respond to aerosol smoke.

Euryops virgineus **(L. f.) Less.**

HONEY EURYOPS

Euryops virgineus is also native to the Cape Floristic Region of South Africa. Aerosol smoke (30 minutes) promoted germination of its seed (Brown et al. 2003; Brown and Botha 2004).

Felicia aethiopica **ssp.** aethiopica **(Burm. f.) Bolus & Wolley-Dod ex Adamson & T.M. Salter**

KINGFISHER DAISY

Germination of kingfisher daisy, a South African fynbos species, was significantly increased by 167% when its seeds were exposed to aerosol smoke for 30 minutes (Brown and Botha 2004). A previous study did not, however, improve germination (Brown et al. 2003).

Felicia amelloides **(L.) Voss**

BLUE FELICIA BUSH

Germination of this species, a native of the Cape Floristic Region of South Africa, was not significantly improved when its seeds were treated for 30 minutes with aerosol smoke (Brown and Botha 2004). The species reported was its synonym, *Felicia aethiopica*.

Felicia filifolia **Burtt Davy**

FINE-LEAVED BUSH DAISY

Germination of this species, also a native of the Cape Floristic Region of South Africa, was not significantly improved when its seeds were treated with 30 minutes of aerosol smoke (Brown and Botha 2004).

Felicia heterophylla **(Cass.) Grau**

FELICIA DAISY

Felicia heterophylla is native to the Cape Floristic Region of South Africa. Aerosol smoke (30 minutes) promoted germination of its seeds (Brown et al. 2003; Brown and Botha 2004).

Gamochaeta subfalcata **(Cabrera) Cabrera**

This species is native to the Americas. Soil samples containing the seeds of this species were collected across forest edges between subtropical rainforest and eucalypt forest in the Lamington National Park of Queensland, Australia, and exposed to cool aerosol smoke for a period of 60 minutes with no significant improvement in germination (Tang et al. 2003).

Garberia heterophylla **(W. Bartram) Merr. & F. Harper**

GARBERIA

Exposure to aerosol smoke for 10 and 30 minutes inhibited seed germination in this upland Florida, United States, plant species (Lindon and Menges 2008).

Gnaphalium californicum **DC.**

CALIFORNIA EVERLASTING

This species commonly occurs west of the Rocky Mountains of the United States. Charred wood (0.5 g on filter paper) significantly improved seed germination (Keeley and Keeley 1987).

Gnephosis tenuissima **Cass.**

Gnephosis tenuissima is commonly found in grasslands and scrublands, and often saline substrates of semiarid and arid regions of Australia. The germination of this species was promoted by treating its seeds with karrikinolide (100 ppb) and smoke water (1:10 dilution) (prepared according to Dixon et al. 1995) (Merritt et al. 2006).

Gutierrezia sarothrae **(Pursh) Britton & Rusby**

BROOM SNAKEWEED

Chou et al. (2012) tested smoke water, heat and combinations of both heat and smoke on the germination percentage and mean germination times of this species. Smoke treatments were applied by soaking seeds for 20 hours in the commercially available Regen 2000® smoke water solution, at concentrations of 1:5, 1:10, or 1:100 (v/v). Seeds were exposed to heat treatments of 50 or 80°C for 5 minutes. *Gutierrezia sarothrae* germination percentage significantly increased when treated with the 1:5 concentration of smoke water, whereas mean germination times were significantly inhibited by 1:10 and 1:5 concentrations of smoke water. The seeds of this species did not respond to heat treatments, nor was there a significant interaction between heat and smoke treatments.

Haplopappus schumannii (Kuntze) G. K. Br. & W. D. Clark

Thirty minutes exposure to cool aerosol smoke significantly inhibited germination in this woody species of the Mediterranean matorral of central Chile (Gómez-González et al. 2008).

Helenium aromaticum (Hook.) H. L. Bailey

MANZANILLA DEL CAMPO

Gómez-González et al. (2011) studied the effects of smoke (10 minutes treatment of cold smoke) and a combination of heat and smoke on the seeds of *H. aromaticum* and reported that neither treatment significantly promoted germination. Heat-shock alone did, however, significantly decrease germination. This species is a native weed that occurs throughout the Mediterranean region of Chile.

Helianthus grosseserratus M. Martens

SAWTOOTH SUNFLOWER

Seeds of sawtooth sunflower, collected from a tallgrass prairie in the Midwest of the United States, did not germinate in response to aerosol smoke treatments of 1, 10, or 60 minutes duration (Jefferson et al. 2007).

Helichrysum dasyanthum (Willd.) Sweet

Germination of this native fynbos species of the Cape Floristic Region of South Africa was not significantly improved when its seeds were exposed to 30 minutes of aerosol smoke (Brown and Botha 2004).

Helichrysum foetidum (L.) Moench

SOUTH AFRICAN PAPERDAISY

Helichrysum foetidum is native to the Cape Floristic Region of South Africa. Aerosol smoke (30 minutes) significantly promoted seed germination in this species (Brown et al. 2003).

Helichrysum pandurifolium Schrank

Germination of this native of the Cape Floristic Region of South Africa was not significantly improved when its seeds were exposed to 30 minutes of aerosol smoke (Brown and Botha 2004).

Helichrysum patulum (L.) D. Don.

HONEY EVERLASTING

Helichrysum foetidum is also native to the Cape Floristic Region of South Africa. Aerosol smoke (30 minutes) significantly promoted germination of its seeds (Brown et al. 2003; Brown et al. 1995; Brown and Botha 2004). Brown and Botha (2004) reported that some samples of this species did not respond to aerosol smoke.

Helichrysum tinctum **(Thunb.) Hilliard & B. L. Burtt.**

Helichrysum tinctum is native to the Cape Floristic Region of South Africa. Aerosol smoke (30 minutes) significantly promoted germination of its seeds (Brown et al. 1995; Brown et al. 2003; Brown and Botha 2004).

Hirpicium alienatum **(Thunb.) Druce**

HAARBOSSIE (AFR.)

Haarbossie occurs in the Cape Floristic Region of South Africa. An application of aerosol smoke did not improve seed germination (Brown et al. 2003).

Hyalosperma cotula **(Benth.) Wilson.**

This annual herb occurs throughout the proteaceous heaths, jarrah forests, and banksia woodlands of Western Australia. Roche et al. (1997a) treated the soil seed bank of a bauxite mine site with smoke water with germination being significantly promoted. Norman et al. (2006) reported similar results when seeds were pre-imbibed in 1% smoke water for 60 minutes prior to sowing them in the soils of a bauxite mine rehabilitation area.

Hymenolepis parviflora **(L.) DC.**

COULTER-BUSH

Coulter-bush occurs in the Cape Floristic Region of South Africa. An application of aerosol smoke did not improve seed germination (Brown et al. 2003).

Hypochaeris radicata **L.**

CATSEAR

Catsear is native to Europe, but also occurs in the Americas, Australia, Japan, and New Zealand. Soil samples containing seeds of this species were collected across forest edges between subtropical rainforest and eucalypt forest in the Lamington National Park of Queensland, Australia, and were exposed for 60 minutes to cool aerosol smoke. There was no improvement in germination following this treatment (Tang et al. 2003).

Ixodia achillaeoides **R. Br.**

MOUNTAIN DAISY

Germination of mountain daisy seeds, collected from a *Eucalyptus baxteri* heathy-woodland in Victoria, Australia, significantly improved following a combined heat and smoke water treatment (Enright and Kintrup 2001).

Lactuca sativa **L. cv.**

LETTUCE

Lactuca sativa is cultivated lettuce. Light et al. (2002) showed that germination was both promoted and inhibited by smoke water dilutions. Seed germination was dependent on the concentration of the smoke water and the duration of soaking (see also Drewes et al. 1995). Karrikinolide significantly increased germination at concentrations ranging from 1 ppm to 1 ppt (Flematti et al. 2004). Germination was equivalent or higher than that obtained by treating the seeds with smoke water (0.01%, 0.1%, 1%, or 10%). Flematti et al. (2007) showed that a number of analogues of karrikinolide were effective in also promoting germination in this species. Jäger et al. (1996) reported that two commercial food flavourants, Smoke Liquid Flavor 94275 and pyroligenous acid ex-eucalyptus wood 621053, both stimulated germination of light-sensitive Grand Rapids lettuce seeds.

Lagenifera huegelii **Benth.**

COARSE BOTTLE DAISY

Aerosol smoke treatments of broadcast seeds in rehabilitated mine sites in the southwest of Western Australia had no effect on their germination (Roche et al. 1997a).

Lagenifera stipitata **(Labill.) Druce**

BLUE BOTTLE DAISY

Soil samples from the Eden Burning Study Area, a dry sclerophyll forest in the Yalumba State Forest of New South Wales, Australia, were collected and air dried to test the effects of heat, smoke, and an interaction between the two cues on seeds from the seed bank. Samples exposed to heat treatment were incubated at 80°C for 60 minutes while those exposed to smoke were incubated in a room, where smoke was generated for 120 minutes. Only the smoke treatment significantly improved germination for this species (Penman et al. 2008). Neither the heat treatment nor the interaction between the two cues had any effect.

Lawrencella davenportii **(F. Muell.) Paul G. Wilson**

DAVENPORT'S DAISY

Davenport's daisy is widely distributed across the semiarid and arid regions of Western Australia, South Australia, and the Northern Territory, Australia. Flematti et al. (2004) reported that germination significantly increased when its seeds were treated with karrikinolide (10 ppb) for a period of 24 hours prior to sowing.

Leucochrysum albicans **(A. Cunn.) Paul G. Wilson ssp.** albicans

HOARY SUNRAY

This subspecies of hoary sunray is common to the grassy woodlands of the New England Tableland of New South Wales, Australia, and elsewhere in eastern

Australia. Smoke water treatments had no significant effect on seed germination (Clarke et al. 2000).

Liatris aspera **Michx.**

TALL BLAZING STAR

Seed of *Liatris aspera*, collected from a tallgrass prairie in the Midwest of the United States, did not germinate in response to aerosol smoke treatments of 1, 10, or 60 minutes duration (Jefferson et al. 2007).

Liatris chapmanii **Torr. & A. Gray**

CHAPMAN'S BLAZING STAR

Exposure to aerosol smoke for 1 and 5 minutes significantly promoted seed germination in this upland Florida plant species (Lindon and Menges 2008).

Liatris mucronata **DC.**

CUSP BLAZING STAR

Aerosol smoke treatments of 8 minutes, in combination with heat treatments of either 30 or 60 seconds in an oven at 100°C and wet, cold stratification for 1 month at 4°C, did not significantly improve germination for this species (Schwilk and Zavala 2012). Stratifying the seeds in a dry, cold environment at 4°C for 1 month, at a relative humidity of 10%, had no effect on germination either.

Liatris ohlingerae **(S. F. Blake) B. L. Rob.**

FLORIDA BLAZING STAR

Exposure to aerosol smoke for 1 minute promoted germination in the seeds of this United States species, while exposures of 10 and 30 minutes inhibited it (Lindon and Menges 2008).

Liatris pycnostachya **Michx.**

PRAIRIE BLAZING STAR

Like *L. mucronata* above, aerosol smoke treatments of 4 or 8 minutes, heat treatments of 30 or 60 seconds at 100°C and either wet, cold stratification for 1 month at 4°C or dry, cold stratification at 4°C for a period of 1 month at a relative humidity of 10%, had no effect on seed germination for this species (Schwilk and Zavala 2012).

Liatris scariosa **(L.) Willd.**

DEVIL'S BITE

Ex situ germination of devil's bite, native to the eastern United States, was promoted when seeds were treated with aerosol smoke for 60 minutes (Jefferson et al. 2007).

Malacothrix clevelandii **A. Gray**

CLEVELAND'S DESERT DANDELION

Seed germination of this chaparral species of California and Arizona responded positively to a powdered charred wood (0.5 g) application (Keeley and Keeley 1987).

Metalasia densa **(Lam.) P. O. Karis**

BLOMBOS (AFR.)

Aerosol smoke applications 30 minutes durations promoted the germination of this South African fynbos species. There was only 1% germination without treatment 14% with it (Brown 1993a; see also Brown et al. 1995; Brown and Botha 2004).

Metalasia muricata **(L.) D. Don**

WHITE BRITTLE BUSH

White brittle bush occurs in the Cape Floristic Region of South Africa. An application of aerosol smoke did not significantly affect germination (Brown et al. 2003).

Microseris lanceolata **(Walp.) Sch. Bip.**

YAM DAISY

Yam daisy is common to the grassy woodlands of the New England Tableland of New South Wales, Australia, and elsewhere in eastern Australia. Smoke water treatments had no effect on the germination of this species (Clarke et al. 2000).

Myriocephalus guerinae **F. Muell.**

Myriocephalus guerinae is an annual herb, which commonly occurs in the semiarid and arid regions of Western Australia. Seed germination was promoted by treating the seeds with karrikinolide (100 ppb) and smoke water (1:10 dilution) (prepared according to Dixon et al. 1995) (Merritt et al. 2006). Gibberellic acid (1,000 ppm) was more effective than karrikinolide (1, 10, or 100 ppb) at promoting germination under light and dark conditions (Merritt et al. 2006).

Oedera capensis **(L.) Druce**

Oedera capensis is a native of South Africa. Brown et al. (2003) reported that by subjecting the seeds to aerosol smoke for 30 minutes, germination increased to 20% in comparison to 5% for the control group.

Olearia elliptica **DC.**

STICKY DAISY BUSH

Sticky daisy bush is common to the grassy woodlands of the New England Tableland of New South Wales, Australia, and elsewhere in eastern Australia. Smoke water treatments had no effect on the germination of this species (Clarke et al. 2000).

Oligoneuron rigidum **(L.) Small**

STIFF GOLDENROD

This species, previously known as *Solidago rigida*, has a wide distribution throughout the eastern and central United States and eastern and central Canada. This species has become a weed in other parts of the world. Germination was significantly promoted when the seeds of this species were exposed to aerosol smoke for 1, 10, or 60 minutes (Jefferson et al. 2007).

Oncosiphon grandiflorum **(Thunb.) Källersjö**

Originally named *Tanacetum grandiflorum*, this species occurs in the Cape Floristic Region of South Africa. An application of aerosol smoke did not improve seed germination for this species (Brown et al. 2003).

Oncosiphon suffruticosum **(L.) Källersjö**

SHRUBBY MAYWEED

Like *O. grandiflorum*, this species of the Cape Floristic Region of South Africa has also recently undergone a name change. The shrubby mayweed was originally called *Tanacetum multiflorum*. An application of aerosol smoke did not improve seed germination for this species (Brown et al. 2003).

Osteospermum fruiticosum **(L.) Norl.**

TRAILING AFRICAN DAISY

The trailing African daisy occurs in the Cape Floristic Region of South Africa and elsewhere in the world, including the state of California in the United States. An application of aerosol smoke did not improve seed germination for this species (Brown et al. 2003). This species is sometimes also referred to as its synonym, *Dimorphotheca fruticosa*.

Othonna bulbosa **L.**

This species occurs in the Cape Floristic Region of South Africa. An application of aerosol smoke did not improve seed germination for this species (Brown et al. 2003).

Othonna parviflora **P. J. Bergius**

Othonna parviflora is native to the Cape Floristic Region of South Africa. Aerosol smoke (30 minutes) significantly promoted germination in this species (Brown et al. 2003).

Othonna quinquedentata **Thunb.**

Germination in this species, a South African fynbos plant, was significantly enhanced following 30 minutes of aerosol smoke and smoke water treatments. The rate of germination also significantly increased (Brown 1993a; Brown et al. 1995; Brown et al. 2003; Brown and Botha 2004).

Ozothamnus **sp.**

Germination of a species of *Ozothamnus*, was significantly promoted by soaking the seeds in a solution with karrikinolide (10 ppb for 24 hrs) (Flematti et al. 2004). This species was recorded as *O. cordifolium* by Flematti et al. (2004), but was not found on the IPNI, APNI, or FloraBase websites or during an Internet search.

Parthenium integrifolium **L.**

WILD QUININE

Wild quinine commonly occurs throughout the eastern and central United States. Germination was promoted when its seeds were treated with aerosol smoke for 1 minute, but was inhibited when seeds were treated with aerosol smoke for 60 minutes (Jefferson et al. 2007).

Phaenocoma prolifera **(L.) D. Don**

CAPE STRAWFLOWER

Phaenocoma prolifera is native to the Cape Floristic Region of South Africa. Aerosol smoke (30 minutes) significantly promoted germination of its seeds (Brown et al. 1995; Brown et al. 2003; Brown and Botha 2004). Brown and Botha (2004) also reported that some samples of this species did not respond to aerosol smoke.

Pityopsis graminifolia **(Michx.) Nutt.**

NARROWLEAF SILKGRASS

Exposure to aerosol smoke for 10 and 30 minutes inhibited seed germination in this upland Florida plant species (Lindon and Menges 2008).

Plecostachys serpyllifolia **(Berg.) Hilliard & B. L. Burtt**

CLIPPED LIME

Clipped lime occurs in the Cape Floristic Region of South Africa and has naturalized in California (Riefner 2009). An application of aerosol smoke did not improve seed germination for this species (Brown et al. 2003).

Podolepis canescens **A.Cunn ex DC.**

LARGE COPPER-WIRE DAISY

Large copper-wire daisy is widely distributed throughout Australia. Karrikinolide (100 ppb) and smoke water (1:10 dilution; prepared according to Dixon et al. 1995) promoted the ex situ germination of *P. canescens* (Merritt et al. 2006). Gibberellic acid (1,000 ppm) was as effective as karrikinolide (1, 10, or 100 ppb) at promoting germination under both light and dark conditions (Merritt *et al.* 2006).

Podolepis gracilis (Lehm.) Graham

SLENDER PODOLEPIS

This species occurs in the jarrah (*Eucalyptus marginata*) forests of Western Australia. Neither aerosol smoke nor smoke water treatments significantly promoted germination of its seeds (Norman et al. 2006).

Podolepis jaceoides (Sims) Voss

SHOWY COPPER-WIRE DAISY

This showy copper-wire daisy is widespread throughout southeastern Australia. Germination was inhibited by 90% when subjected to a 10% smoke water application (Clarke et al. 2000).

Podolepis lessonii (Cass.) Benth.

YELLOW BUTTONS

This species occurs in the jarrah (*Eucalyptus marginata*) forests of Western Australia. Neither aerosol smoke nor smoke water treatments significantly promoted germination of its seeds (Norman et al. 2006).

Podolepis monticola R. J. F. Hend.

Podolepis monticola was first described from samples found in Queensland, Australia. Soil samples containing the seeds of this species were collected across forest edges between subtropical rainforests and eucalypt forests in the Lamington National Park of Queensland, Australia, and treated with cool aerosol smoke for a period of 60 minutes. There was no significant improvement to germination (Tang et al. 2003).

Pterochaeta paniculata Steetz

Aerosol smoke treatments of broadcast seeds in rehabilitated mine sites in the southwest of Western Australia had no significant effect on their germination (Roche et al. 1997a).

Rafinesquia californica Nutt.

CALIFORNIA CHICORY

California chicory, which occurs throughout the western United States, exhibited significantly greater germination following treatment with charate (0.5 g on filter paper on which the seeds were placed). Germination increased from 4% in the control (no treatment) to 55% when treated with charate (Keeley and Keeley 1987).

Rhodanthe citrina (Benth.) Paul G. Wilson

Rhodanthe citrina commonly occurs in the Southwest Botanical Province of Western Australia, but can also be found growing in the semiarid and arid regions of

Australia. Karrikinolide (100 ppb) and smoke water (1:10 dilution; prepared according to Dixon et al. 1995) treatment of seeds promoted the ex situ germination of *R. citrina* (Merritt et al. 2006).

Rudbeckia hirta **L.**

BLACK-EYED SUSAN

Seed of black-eyed Susan, collected from a tallgrass prairie in the Midwest of the United States, did not germinate in response to aerosol smoke treatments of 1, 10, or 60 minutes duration (Jefferson et al. 2007).

Schoenia cassiniana **Steetz**

PINK EVERLASTING DAISY

This annual daisy has a wide distribution across semiarid and arid environments of Australia. Ex situ application of aerosol smoke to sown seeds had a significant negative effect on final germination percentages (Roche et al. 1997b).

Senecio bracteolatus **Hook. f.**

SENECIO

The germination of *S. bracteolatus* seeds, when treated with a combination of heat (80°C for 5 minutes) and aerosol smoke for 60 minutes, was significantly inhibited (Gonzalez and Ghermandi 2012). This species is native to the grasslands of northwestern Patagonian.

Senecio grandiflorus **P. J. Bergius**

BIG-LEAF GROUNDSEL

Big-leaf groundsel is native to the South African fynbos communities. Brown (1993a) reported that germination was promoted following 30 minutes of aerosol smoke treatment (see also Brown et al. 1995). Smoke water did not increase the total number of germinants over the range of dilutions tested; however, the rate of germination increased. A 1:10 dilution had an inhibitory effect on seed germination.

Senecio halmifolius **L.**

Seed germination in this species of the Cape Floristic Region of South Africa was not significantly improved following 30 minutes of aerosol smoke treatment (Brown et al. 2003; Brown and Botha 2004).

Senecio jacobaea **L.**

TANSY RAGWORT

Tansy ragwort is native to southern Europe, northern Africa and western Asia, and has become a weed in many parts of the world. Pérez-Fernández and Rodríguez-Echeverría (2003) reported that germination of this species was significantly promoted when its seeds were exposed to aerosol smoke for 10 minutes. Germination was also promoted when the seeds were treated with karrikinolide (10^{-7} M; Daws et al. 2007).

Senecio pinifolius **Lam.**

Senecio pinifolius occurs in the Cape Floristic Region of South Africa. An application of aerosol smoke did not improve seed germination in this species (Brown et al. 2003).

Senecio rigidus **L.**

Brown (1993a) showed that incubation of seeds in a smoke water extract (1:2 dilution) significantly inhibited germination (80% germination in the control versus 30% germination in smoke water treatment). Brown et al. (2003) have since revealed that germination of this South African species is also promoted by aerosol smoke (see also Brown and Botha 2004).

Senecio umbellatus **L.**

Senecio umbellatus is native to the Cape Floristic Region of South Africa. Aerosol smoke (30 minutes) significantly promoted germination of its seeds (Brown et al. 2003; Brown and Botha 2004). Brown et al. (2003), and Brown and Botha (2004) also reported that some samples of this species did not respond to aerosol smoke.

Sigesbeckia orientalis **L.**

INDIAN WEED

Soil samples containing seeds of this exotic species, which occurs in several places in the world, were collected across forest edges between subtropical rainforest and eucalypt forest in the Lamington National Park of Queensland, Australia, and were treated with cool aerosol smoke for a period of 60 minutes with no significant effect in germination (Tang et al. 2003).

Silphium laciniatum **L.**

COMPASS PLANT

Compass plant seed, collected from a tallgrass prairie in the Midwest of the United States, did not germinate in response to aerosol smoke treatments of 1, 10, or 60 minutes duration (Jefferson et al. 2007).

Solidago virgaurea **L. ssp.** asiatica **Kitam. ex. Hara.**

GOLDENROD

The effects of aerosol smoke, heat, darkness, cold stratification, and combinations of smoke with each of the three other treatments on seed germination were examined in this study (Tsuyuzaki and Miyoshi 2009). Smoke was produced by burning Timothy hay (*Phleum pratense*), which was pumped through a 3.5 m cooling tube into a smoke chamber for approximately 5 minutes. The seeds were exposed to the smoke for 60 minutes. Those seeds exposed also to heat were incubated at 75°C for 25 minutes. The cold stratification process took 1 month, during which the seeds remained in an incubator set to 4°C. Where the dark treatment was concerned, the seeds were maintained in total darkness for the entire germination period. Germination percentage was recorded for each treatment. Two treatments significantly inhibited the germination of this species: smoke (10% germination), and dark (17% germination) treatments. This was compared to 92% germination for the control group.

Stephanomeria virgata **Benth.**

ROD-WIRE LETTUCE

Rod-wire lettuce is native to California, Nevada, and Oregon, United States. Germination was promoted by an application of charate (0.5 g per Petri dish) (Keeley and Keeley 1987).

Stipa speciosa **Trin. & Rupr.**

DESERT NEEDLEGRASS

The seeds of this native species of the grasslands of northwestern Patagonian were treated with a combination of heat (80°C for 5 minutes) and were exposed to aerosol smoke for 60 minutes with no significant effect to germination (Gonzalez and Ghermandi 2012).

Stuartina muelleri **Sond.**

SPOON CUDWEED

Germination of spoon cudweed seed, collected from a *Eucalyptus baxteri* heathy-woodland in Victoria, Australia, did not significantly improve following a 24-hour smoke water treatment (Enright and Kintrup 2001).

Symphyotrichum falcatum **(Lindl.) G. L. Nesom**

WHITE PRAIRIE ASTER

The seeds of this species were collected from a frequently burned ponderosa pine forest (*Pinus ponderosa*) in northern Arizona, and were treated with concentrated aqueous smoke (Wright's Brand, Roseland, New Jersey). This significantly increased the percentage of emergence of seedlings (Abella and Springer 2009).

Symphyotrichum laeve **(L.) A. Löve & D. Löve var.** leave

SMOOTH BLUE ASTER

The seeds of this variety of smooth blue aster, collected from a tallgrass prairie in the Midwest of the United States, did not germinate in response to aerosol smoke treatments of 1, 10, or 60 minutes duration (Jefferson et al. 2007).

Syncarpha eximia **(L.) B. Nord.**

STRAWBERRY EVERLASTING

Syncarpha eximia (native to the fynbos in South Africa) achieved a mean germination of 80% when sown seeds were exposed to aerosol smoke for 30 minutes. That was compared to 19% for the control (no treatment) (Brown 1993a; Brown et al. 1995; Brown and Botha 2004). Smoke extracts had no effect on final germination percentage, but did increase the rate of germination (Brown 1993a). In another study, there was no significant response in germination to aerosol smoke treatments (Brown et al. 2003).

Syncarpha speciosissima **(L.) B. Nord.**

Syncarpha speciosissima is native to the Cape Floristic Region of South Africa. An aerosol smoke treatment of 30 minutes significantly promoted seed germination (Brown et al. 1995; Brown et al. 2003; Brown and Botha 2004).

Syncarpha vestita **(L.) B. Nord.**

TINDER EVERLASTING

For this native of the South African fynbos, germination increased significantly in response to exposing sown seeds to aerosol smoke for 30 minutes. That was in comparison to no treatment (78% and 2% final germination, respectively) (Brown 1993b; Brown and Botha 2004). Smoke extracts were also effective in promoting germination for all of the dilutions tested (Brown 1993b). Pretreatment in smoke solution broke dormancy in the seeds of this species and significantly improved germination (Brown et al. 1998). Seeds pretreated in this manner and later stored at 18°C for periods of up to 12 months exhibited similar results. Flematti et al. (2004) achieved 78% germination by treating the seeds with karrikinolide (10 ppb).

Trichocline spathulata **(DC.) J. H. Willis**

NATIVE GERBERA

This species occurs in the jarrah (*Eucalyptus marginata*) forests of Western Australia. Neither aerosol smoke nor smoke water treatments significantly promoted germination of its seed (Roche et al. 1997a,b; Norman et al. 2006).

Tripteris sinuata **DC.**

KAROO BIETOU (AFR.)

Karoo bietou occurs in the Cape Floristic Region of South Africa. Treatments with aerosol smoke did not improve seed germination for this species (Brown et al. 2003).

Ursinia paleacea **(L.) Moench.**

This species is native to the Cape Floristic Region of South Africa. Aerosol smoke (30 minutes) was shown to promote germination (Brown et al. 1995; Brown et al. 2003; Brown and Botha 2004).

Ursinia sericea **(Thunb.) N. E. Br.**

LACE-LEAF URSINIA

Lace-leaf ursinia occurs in South Africa's Cape Floristic Region and succulent Karro Biomes. Aerosol smoke did not improve seed germination for this species (Brown et al. 2003).

Ursinia tenuifolia **(L.) Poir.**

This species occurs in South Africa's Cape Floristic Region. An application of aerosol smoke did not improve seed germination for this species (Brown et al. 2003).

Vittadinia gracilis **(Hook. f.) N. T. Burb.**

WOOLLY NEW HOLLAND DAISY

Common to the southern parts of the east coast of Australia, the freshly collected seeds of this species, when treated for 60 minutes with cool aerosol smoke, did not germinate (Roche et al. 1997b).

Vittadinia muelleri **N. T. Burb.**

NARROW-LEAF NEW HOLLAND DAISY

Vittadinia muelleri is common to the grassy woodlands of the New England Tableland of New South Wales, Australia, and elsewhere in eastern Australia. Smoke water treatments had no effect on the germination of this species (Clarke et al. 2000).

BETULACEAE

Alnus glutinosa **(L.) Gaertn.**

EUROPEAN ALDER

Crosti et al. (2006) reported that exposure to cool aerosol smoke for for a 60 minute period significantly promoted germination by 60% in this species, which commonly occurs in the fire-prone habitats of Europe.

Alnus maximowiczii **Callier ex. C. K. Schneid.**

MAXIMOWICZ ALDER

The effects of aerosol smoke, heat, darkness, cold stratification, and combinations of smoke with each of the three other treatments on seed germination were examined in this study (Tsuyuzaki and Miyoshi 2009). Smoke was produced by burning Timothy hay (*Phleum pratense*), which was pumped through a 3.5 m cooling tube into a smoke chamber for approximately 5 minutes. The seeds were exposed to the smoke for 60 minutes. Those seeds exposed also to heat were incubated at 75°C for 25 minutes. The cold stratification process took 1 month, during which the seeds remained in an incubator set to 4°C. Where the dark treatment was concerned, the seeds were maintained in total darkness for the entire germination period. The dark treatment, when tested on its own, significantly inhibited the germination of the seeds of this species (0% dark versus 74% control). None of the other treatments had any effect on germination.

Betula platyphylla **Sukaczev var.** japonica **(Miq.) Hara**

JAPANESE WHITE BIRCH

The effects of aerosol smoke, heat, darkness, cold stratification, and combinations of smoke with each of the other three treatments on seed germination were examined in this study (Tsuyuzaki and Miyoshi, 2009; see *Alnus maximowiczii* above for details of the tests performed). The dark treatment, when tested on its own, as well as the combination of aerosol smoke and cold stratification treatments significantly inhibited germination, with 1% and 12% germination, respectively, compared to the control group (38%). None of the other treatments had any effect on germination. The seeds of this species display physiological dormancy.

BIGNONIACEAE

Eccremocarpus scaber **Ruiz & Pav.**

CHILEAN GLORY FLOWER

Thirty minutes exposure to cool aerosol smoke significantly inhibited germination in this woody species of the Mediterranean matorral of central Chile (Gómez-González et al. 2008).

BORAGINACEAE

Cryptantha clevelandii **Greene**

WHITE FORGET-ME-NOT

Cryptantha clevelandii is native to the Californian chaparral. Keeley and Fotheringham (1998a) reported that germination of this species increased in response to aerosol smoke. This species is considered vulnerable globally and is likely to be moved to endangered status if the factors causing species decline continue (IUCN 2013).

Cryptantha intermedia **(Gray) Greene**

LARGE-FLOWERED POPCORN FLOWER

Cryptantha intermedi is widely distributed throughout western Canada, western United States and northern Mexico. Keeley and Keeley (1987) reported that charate (0.5 g per Petri dish) significantly enhanced its germination.

Cryptantha micrantha **(Torr.) I. M. Johnst.**

RED-ROOT CRYPTANTHA

Like *C. clevelandii*, this species is also a native of the Californian chaparral of the United States. Keeley and Fotheringham (1998a) reported that germination of its seeds increased in response to aerosol smoke treatments.

Cryptantha muricata **(Hook. & Arn.) A. Nelson & J. F. Macbr.**

PRICKLY CRYPTANTHA

Native to the chaparral of California, Nevada, and Arizona, germination of this annual was promoted when its seeds were treated with powdered charred wood (Keeley et al. 1985). The IUCN (2013) has ranked this species as rare and at risk globally.

Cynoglossum australe **R. Br.**

AUSTRALIAN HOUND'S TONGUE

Australian hound's tongue is common to the grassy woodlands of the New England Tableland of New South Wales, Australia, and elsewhere in eastern Australia. Smoke water treatments had no effect on the germination of this species (Clarke et al. 2000).

Echium plantagineum **L.**

PATERSON'S CURSE

Native to most of Europe, western Asia and northern Africa, Paterson's curse has naturalized in Australia, the United States, South America, New Zealand, and South Africa. Germination increased significantly in response to smoke water (10:10 v/v) and karrikinolide (0.67μM) treatments (Stevens et al. 2007).

BRASSICACEAE

Arabidopsis thaliana **(L.) Heynh**

MOUSE EAR CRESS

Pennacchio et al. (2007b) reported that rate of germination in the seeds of the Col-3 ecotype of this species was significantly increased following aerosol smoke treatments. Prolonged exposures to smoke, in contrast, significantly inhibited germination. The discovery of smoke-derived karrikinolide-like substances, the karrikins, have since been shown to trigger germination in the seeds of *A. thaliana* via a mechanism that requires both light and gibberellic acid synthesis (Nelson et al. 2009).

Arabis fendleri **Greene**

FENDLER ROCKCRESS

The seeds of this species, collected from a ponderosa pine (*Pinus ponderosa*) forest in northern Arizona did not significantly respond to concentrated aqueous smoke treatment (Wright's Brand, Roseland, New Jersey) (Abella and Springer 2009).

Brachycarpaea juncea **(P. J. Berg.) Marais**

Seed germination in this fynbos species of the Cape Floristic Region of South Africa was not significantly improved following 30 minutes of aerosol smoke treatment

(Brown and Botha 2004). The researchers note, however, that the species has an impenetrable seed coat.

Brassica oleracea **L. var. Italica**

BROCCOLI

A combination of aerosol smoke treatments of 15, 30, or 45 minutes, in combination with three different treatments of dissolved aspirin (0.145 wt%; 10, 20, and 30 minutes; Bayer) increased both germination of broccoli seeds and their growth ratio (Hong and Kang, 2011).

Brassica napus **L.**

RAPESEED

Abdollahi et al. (2012) tested an aqueous smoke extract, prepared by burning the leaves of the medicinal plant, feverfew (*Tanacetum parthenium* L.). They pumped the smoke through a column of distilled water (500 mL) for 45 min, as according to the method described by Baxter et al. (1994). A neutral fraction that excluded both weak and strong acids, as well as phenols, was then obtained according to the method described by Flematti et al. (2007). From this fraction, the following concentrations (v/v) of smoke extract were prepared: 1:10, 1:100, 1:250, 1:500, 1:1,000, and 1:2,000. Applying the smoke extracts, before or during the germination of microspore-derived embryos (MDEs), significantly improved secondary embryogenesis in *Brassica napus*. Incubating MDEs for less than 15 minutes with the 1:250 smoke extract yielded the best results.

Brassica napus **L. cv. Apex**

RAPESEED

Rapeseed is cultivated in many parts of the world and is prized for its oil and use in agriculture as an animal feed. Thornton et al. (1999) reported that dormancy, induced by osmotic stress using polyethylene glycol (PEG 8000), could be broken if the seeds were immersed in a solution of smoke made with burnt straw of *Triticum aestivum*.

Brassica napus **L. cv. Topas**

RAPESEED

Inoculation of canola microspore-derived embryos with aqueous smoke dilutions of 1:250, 1:500, 1:1,000, and 1:2,000 for 5, 10, and 15 minute periods significantly increased the percentage of plantlet regeneration from the embryos while decreasing callogenesis (Ghazanfari et al. 2012). Significantly greater plantlet regeneration was achieved when the smoke solutions were applied to embryos in a B_5 regeneration medium, with the best results occurring following treatment with smoke dilutions of 1:250. Root and shoot lengths were also significantly increased, but not when treated with additional dilutions of 1:10 and 1:100. When used in combination with gibberellic acid (0.1 and 0.15 mg/L), the less concentrated dilutions further improved root and shoot lengths. The conversion frequencies of rapeseed microspore-derived embryos could also be significantly increased when pretreated with smoke water solutions. Rapeseed, also known as canola, is one of the largest edible vegetable oil crops produced in the world.

Brassica nigra **(L.) Koch.**

BLACK MUSTARD

The seeds of this species, which is native to Eurasia, are used to make mustard and thus have been intentionally grown in the United States for that purpose. Black mustard is now considered an invasive species in the United States. Keeley et al. (1985) discovered that seed germination was inhibited by charate.

Brassica tournefortii **Gouan**

ASIAN MUSTARD

This mustard is native to southern Europe, northern Africa, temperate Asia and Pakistan. Asian mustard has naturalized in the southwestern United States, Australia, New Zealand, and the British Isles. Stevens et al. (2007) reported that smoke water (10:10 v/v) and karrikinolide (0.67 µM) significantly improved germination for this species when its seeds were incubated for a 12/12-h diurnal temperature regime of 18/5°C. Germination increased from less than 5% (control) to approximately 60% and 95% when treated with smoke water and karrikinolide, respectively. In another study by Long et al. (2011), freshly collected *B. tournefortii* seeds were tested for the following effects to karrikinolide (KAR1) and combinations with several fire cues: Test 1, the effects on germination when 1 µM KAR_1 was tested over a range of temperature and light treatments; Test 2, response to KAR_1 after various dormancy breaking treatments were implemented, wet-dry cycling or dry after-ripening combinations from 1 to 3 months; Test 3, responses to KAR_1 following cold (5°C) and warm (20°C) stratification. This was for up to 3 months; and Test 4, the response to KAR_1 after the seeds were buried for up to 2 years. This last test was performed on seed lots collected at three different locations. The results of Test 1 revealed that the highest germination occurred after 21 days, when the seeds were incubated in constant darkness at temperatures ranging between 10 and 20°C. The seeds used in Test 2 continued to be responsive to KAR_1 and to all the treatments tested, however, wet-dry cycling for 2–3 months significantly inhibited germination. There was a significant increase when KAR_1-treated seeds were subject to a warm stratification treatment of 4 to 8 weeks (Test 3). The germination response to KAR_1 in Test 4 varied between sites, with the lowest germination being recorded for seeds buried at the wettest burial site. Germination for freshly collected seeds, or following burial in the first few months, increased significantly, but decreased in the months that followed.

Capsella bursa-pastoris **(L.) Medik.**

SHEPHERD'S PURSE

Seed germination of this global weed was significantly promoted when its seeds were soaked in smoke water (Capeseed, Cape Town, South Africa) for 24 hrs (Daws et al. 2007).

Carrichtera annua **(L.) DC.**

WARD'S WEED

Karrikinolide (KAR$_1$) was tested for its effects on germination when applied at 1 μM and under a range of temperature and light combinations. The highest germination percentages were achieved when the seeds were incubated with KAR$_1$ in an alternating light cycle of 12 hours at temperatures below 20°C (Long et al. 2011).

Caulanthus heterophyllus **(Nutt.) Payson**

SAN DIEGO WILD CABBAGE

Caulanthus heterophyllus is synonymous with *Streptanthus heterophyllus* and is native to the chaparral of California, USA. Only 1% of San Diego wild cabbage seeds germinated without treatment. However, application of 0.5 g of charate per Petri dish resulted in 24% germination (Keeley and Keeley 1987). *Caulanthus heterophyllus* germination increased to 75% when the seeds were exposed to charred wood, and for 5, 8, or 15 minutes of aerosol smoke in later studies. Treatment of seeds with smoke water dilutions also exhibited significant increases in germination of this species (Keeley and Fotheringham 1998a, Fotheringham *et al.* 1995). Germination of 97% was achieved when the seeds were treated with karrikinolide (10 ppb) in comparison to 15% for the control group (Flematti et al. 2004).

Guillenia lasiophylla **(Hook. & Arn.) Greene**

CALIFORNIA MUSTARD

Guillenia lasiophylla is widely distributed throughout northern Mexico, the western United States and western Canada. Germination of this species was inhibited by aerosol smoke (5 or 8 minutes) or heat shock treatment of the seeds (Keeley and Fotheringham 1998a). Fotheringham et al. (1995) previously reported, however, that smoke promoted the germination of this species.

Heliophila coronopifolia **L.**

WILD FLAX

Seed germination in this fynbos species of the Cape Floristic Region of South Africa was not significantly improved following 30 minutes of aerosol smoke treatment (Brown et al. 2003; Brown and Botha 2004). The researchers note, however, that the species germinates without the need for smoke treatments.

Heliophila macowaniana **Schltr.**

This species occurs in South Africa's Cape Floristic Region and other parts of the country. An application of aerosol smoke did not improve seed germination for this species (Brown et al. 2003).

Heliophila pinnata **L. f.**

Seed germination in this fynbos species of the Cape Floristic Region of South Africa was not significantly improved following 30 minutes of aerosol smoke treatment (Brown et al. 2003; Brown and Botha 2004).

Heliophila pusilla **L. f.**

HELIOPHILA

In a study by Long et al. (2011), freshly collected *H. pusilla* seeds were tested for the following effects to karrikinolide (KAR_1) and combinations with several fire cues: Test 1, the effects on germination when 1 μM KAR_1 was tested over a range of temperature and light treatments; Test 2, response to KAR_1 after various dormancy breaking treatments were implemented, wet-dry cycling or dry after-ripening combinations from 1 to 3 months; Test 3, responses to KAR_1 following cold (5°C) and warm (20°C) stratification. This was for up to 3 months. The results of Test 1 revealed that *H. pusilla* seeds did not initially germinate due to any combination of temperature, light and KAR_1 treatments. However, in Test 2, after either a 1 month period of dry after-ripening or wet—dry cycling, seed germination was significantly improved when KAR_1 was administered to seeds incubated in complete darkness. Germination was also significantly increased if KAR_1 was administered after warm stratification (Test 3).

Heliophila **sp.**

Seed germination in an unidentified species of *Heliophila*, which occurs in the Cape Floristic Region of South Africa, was significantly promoted following a 30-minute treatment with aerosol smoke (Brown et al. 2003).

Lepidium africanum **(Burm. f.) DC.**

AFRICAN PEPPERWORT

Freshly collected seeds of this species were tested for their germination effects to 1 μM karrikinolide (KAR_1) under a range of temperature and light combinations. Germination increased by up to 95% in response to KAR_1 when seeds were incubated in darkness at 15°C. However, greater germination percentages were achieved when the seeds were incubated in alternating light with optimal temperatures being under 20°C. This species is native to southern Africa.

Lepidium nitidum **Nutt.**

SHINING PEPPERWEED

Shining pepperweed commonly occurs throughout the western United States and northern Mexico. Charate promoted germination of this species. Germination was 2% for the control (no treatment) versus 22% germination for the charate treatment (Keeley and Keeley 1987).

Lepidium virginicum **L.**

VIRGINIA PEPPERWEED

Lepidium virginicum is widely distributed throughout northern and southern America and has become a weed in many parts of the world. In situ germination of its seeds was inhibited by the application of concentrated smoke water (2 L/m^2) to the soil seed bank of a gravel prairie (Jefferson, unpublished data).

Raphanus raphanistrum **L.**

WILD RADISH

This widely naturalized weed has a broad native distribution throughout Europe, northern Africa and parts of Asia. Germination of wild radish increased significantly in response to smoke water (10:10 v/v) and karrikinolide (0.67 µM) treatments (Stevens et al. 2007). In another study by Long et al. (2011), freshly collected wild radish seeds were tested for the following effects to karrikinolide (KAR$_1$) and combinations with several fire cues: Test 1, the effects on germination when 1 µM KAR$_1$ was tested over a range of temperature and light treatments; Test 2, response to KAR$_1$ after various dormancy breaking treatments were implemented, wet-dry cycling or dry after-ripening combinations from 1 to 3 months; Test 3, responses to KAR$_1$ following cold (5°C) and warm (20°C) stratification. This was for up to 3 months; and Test 4, the response to KAR$_1$ after the seeds were buried for up to 2 years. This last test was performed on seed lots collected at three different locations. The results of Test 1 revealed that germination was significantly promoted in wild radish seeds (approximately 90% germination) when KAR$_1$ was applied to seeds incubated in constant darkness and at temperatures up to 20°C. Germination was significantly decreased when KAR$_1$ was administered to seeds that underwent a wet-dry cycling treatment that lasted 1 month (Test 2). Warm stratification treatment ranging from 4 to 8 weeks significantly improved germination (Test 3). The germination response to KAR$_1$ in Test 4 varied between sites, with the lowest germination being recorded for seeds buried at the wettest burial site. Germination for freshly collected seeds, or following burial in the first few months, increased significantly, but decreased in the months that followed.

Rapistrum rugosum **(L.) All.**

COMMON GIANT MUSTARD

In a study by Long *et al.* (2011), freshly collected *R. rugosum* seeds were tested for the following effects to karrikinolide (KAR$_1$) and combinations with several fire cues: Test 1, the effects on germination when 1 µM KAR$_1$ was tested over a range of temperature and light treatments; Test 2, response to KAR$_1$ after various dormancy breaking treatments were implemented, wet-dry cycling or dry after-ripening combinations from 1 to 3 months; Test 3, responses to KAR$_1$ following cold (5°C) and warm (20°C) stratification. This was for up to 3 months. The results for Test 1 revealed that the germination of fresh seeds was significantly promoted by KAR_1, but remained low (<20%) regardless of temperature or light treatment. Where Test 2 was concerned, the seeds of this species did not germinate in response to wet—dry cycling or dry after-ripening, and germinated only to a maximum of 20% for all of the other treatments. In Test 3, germination significantly decreased in response to KAR_1 and cold stratification treatment. This species is native to Eurasia and parts of Africa and has been introduced to other parts of the world.

Rorippa **sp.**

Soil samples containing seeds of an unidentified species of *Rorippa* were collected across forest edges between subtropical rainforest and eucalypt forest in the Lamington National Park of Queensland, Australia, and treated with cool aerosol smoke for a period of 60 minutes. There was no significant effect on germination (Tang et al. 2003).

Sinapis alba **L.**

WHITE MUSTARD

This global weed is thought to be native to the Mediterranean region of Europe. Germination was promoted when its seeds were treated with karrikinolide (10^{-7} M; Daws et al. 2007).

Sinapis arvensis **L.**

CHARLOCK

Charlock is indigenous to Europe and has become invasive in other parts of the world. Germination was inhibited by smoke water treatment (Adkins and Peters 2001).

Sisymbrium erysimoides **Desf.**

MEDITERRANEAN ROCKET

Freshly collected *S. erysimoides* seeds were tested for the following effects to karrikinolide (KAR_1) and combinations with several fire cues (Long et al. 2011): Test 1, the effects on germination when 1 μM KAR_1 was tested over a range of temperature and light treatments; Test 2, response to KAR_1 after various dormancy breaking treatments were implemented, wet-dry cycling or dry after-ripening combinations from 1 to 3 months; Test 3, responses to KAR_1 following cold (5°C) and warm (20°C) stratification. This was for up to 3 months; and Test 4, the response to KAR_1 after the seeds were buried for up to 2 years. This last test was performed on seed lots collected at three different locations. Where Test 1 was concerned, the greatest germination percentage was achieved when the seeds were incubated with KAR_1 in an alternating light cycle of 12 hours and at temperatures between 20 and 30°C. Germination was significantly promoted for all the treatments of Test 2. The greatest germination percentages for seeds used in Test 3 occurred when they were treated with 1 μM KAR₁ after warm stratification. Germination decreased in response when KAR₁ was administered after cold stratification. The seeds were not initially responsive to KAR₁ in Test 4, but did germinate significantly following 2 months of burial. The greatest germination occurred during the winter months. This species is native to Europe, Asia, and northern Africa.

Sisymbrium orientale **L.**

INDIAN HEDGE MUSTARD

The native range of this mustard is not known. It has naturalized throughout southern Europe, northern Africa, the southern and western United States, western Asia, the Arabian Peninsula, India and Pakistan, and other parts of the world. Charate inhibited germination of this species (Keeley and Keeley 1987), as did smoke water (1/10 v/v; Stevens et al. 2007). Germination did, however, increase significantly when its seeds were treated with karrikinolide (0.67 μM; Stevens et al. 2007). In another study by Long et al. (2011), freshly collected seeds of this species were tested for their effects to 1 μM karrikinolide (KAR₁) and various combinations to fire-related cues (Long et al. 2011). There were four tests in all (see *S. orientale* for details). Where Test 1 was concerned, germination was best promoted when the seeds treated with KAR₁ were incubated in alternating light/dark conditions at 20–30°C. The germination response to KAR₁ for the seeds used in Test 2 was variable, but significantly improved when the seeds were tested combination with 1 month of dry after ripening and dark, 2 months of dry after ripening combined with 1 month of wet-dry cycling in the dark, and 1 month of dry after ripening combined with 2 months wet-dry cycling period in the dark. The greatest germination percentage achieved for this species occurred when the seeds were incubated without KAR₁ and were kept in the dark following a wet-dry cycling period of 1 month. Test 3 revealed there was no significant germination response to KAR₁ following cold or warm stratification. The seeds were not initially responsive to KAR₁ in Test 4, but did germinate significantly following 2 months of burial. The greatest germination occurred during the winter months.

BRUNIACEAE

Audouinia capitata **(L.) Brongn.**

FALSE HEATH

This native of the Cape Floristic Region of South Africa was the first species shown by de Lange and Boucher (1990) to germinate more readily when the seeds were treated with smoke (Brown et al. 1995). The IUCN (2013) lists this species as locally and globally vulnerable.

Berzelia abrotanoides (L.) Brongn.

ABROTAN

Seed germination in this fynbos species of the Cape Floristic Region of South Africa was not significantly improved following 30 minutes of aerosol smoke treatment (Brown and Botha 2004). The researchers note, however, that it was probable the seeds were not viable or may have required other dormancy-breaking treatments.

Berzelia galpinii Pillans

Seed germination in this fynbos species of the Cape Floristic Region of South Africa was not significantly improved after 30 minutes of exposure to aerosol smoke (Brown and Botha 2004).

Berzelia lanuginosa Brongn.

VLEIKOLKOL

Berzelia lanuginosa is native to South Africa. Its germination improved significantly when its seeds were incubated in smoke water (1:50 dilution). Germination, however, remained low regardless of treatment. A mean average of 2% germinated in the control and 12% for smoke treated seeds (Brown 1993a). Germination was significantly promoted by exposing the seeds to 60 minutes of aerosol smoke (Brown et al. 1995), but there was no significant effect after only 30 minutes of treatment (Brown and Botha 2004). The researchers note, however, that it was probable the seeds they tested may not have been viable. Brown et al. (2003) reported no significant effect on germination following aerosol smoke treatment.

Brunia albiflora E. Phillips

WHITE-FLOWERED BRUNIA

White-flowered brunia is native to the Cape Floristic Region of South Africa. The germination of its seeds was enhanced by aerosol smoke treatments (Brown et al. 2003; Brown and Botha 2004). Germination percentages remained low, however.

Brunia laevis Thunb.

COFFEE BUSH

Like *B. albiflora*, coffee bush is native to the Cape Floristic Region of South Africa. The germination of its seeds was enhanced by aerosol smoke treatments (Brown et al. 2003; Brown and Botha 2004).

Staavia radiata Dahl

GLASS EYES

Seed germination in this fynbos species of the Cape Floristic Region of South Africa was not significantly improved following 30 minutes of aerosol smoke treatment

(Brown and Botha 2004). The researchers note, however, that it was probable the seeds were not viable or may have required other dormancy-breaking treatments.

CAMPANULACEAE

Campanula lasiocarpa **Cham.**

ALPINE BELLFLOWER

The effects of aerosol smoke, heat, darkness, cold stratification, and combinations of smoke with each of the three other treatments on seed germination were examined in this study (Tsuyuzaki and Miyoshi 2009). Smoke was produced by burning Timothy hay (*Phleum pratense*), which was pumped through a 3.5 m cooling tube into a smoke chamber for approximately 5 minutes. The seeds were exposed to the smoke for 60 minutes. Those seeds exposed also to heat were incubated at 75°C for 25 minutes. The cold stratification process took 1 month, during which the seeds remained in an incubator set to 4°C. Where the dark treatment was concerned, the seeds were maintained in total darkness for the entire germination period. The aerosol smoke and dark and cold treatments, when tested on their own, all significantly inhibited germination in this species. None of the other treatments had any effect.

Cyphia incisa **Willd.**

This species occurs in South Africa's Cape Floristic Region. An application of aerosol smoke did not affect seed germination for this species (Brown et al. 2003).

Isotoma hypocrateriformis **(R. Br.) Druce**

WOODBRIDGE POISON

Woodbridge poison occurs in the jarrah forest ecosystems in the southwestern region of Western Australia. Neither aerosol smoke nor smoke water treatments significantly promoted germination of its seeds (Roche et al. 1997a; Norman et al. 2006).

Lobelia andrewsii **Lammers**

TRAILING LOBELIA

Soil samples containing seeds of this species were collected across forest edges between subtropical rainforests and eucalypt forests in the Lamington National Park of Queensland, Australia, and were exposed for 60 minutes to cool aerosol smoke with no significant effect on seed germination (Tang et al. 2003).

Lobelia coronopifolia **L.**

BUCKS-HORN LOBELIA

Bucks-horn lobelia is native to the Cape Floristic Region of South Africa. Germination of its seeds was increased by aerosol smoke (Brown et al. 2003; Brown and Botha 2004). The germination percentages were low, however.

Lobelia linearis **Thunb.**

This lobelia is also native to the Cape Floristic Region of South Africa. Germination of its seeds was also increased following an aerosol smoke treatment (Brown et al. 2003; Brown and Botha 2004).

Lobelia pinifolia **L.**

Seed germination of this native of the Cape Floristic Region of South Africa was not significantly improved following 30 minutes of aerosol smoke treatment (Brown and Botha 2004). These researchers have noted, however, that the seed tested may have been of poor viability or may require special dormancy-breaking treatments.

Lobelia rhombifolia **de Vriese**

TUFTED LOBELIA

Aerosol smoke treatments of broadcast tufted lobelia seeds used in rehabilitated mine sites in the southwest of Western Australia did not significantly improve germination (Roche et al. 1997a).

Lobelia **sp.**

Germination of an unidentified species of *Lobelia*, which occurs in the Cape Floristic Region of South Africa, was significantly increased when treated with aerosol smoke (Brown et al. 2003; Brown and Botha 2004).

Monopsis lutea **(L.) Urb.**

YELLOW LOBELIA

Yellow lobelia is native to the Cape Floristic Region of South Africa. Germination of its seeds was significantly increased following an aerosol smoke treatment (Brown et al. 2003; Brown and Botha 2004).

Roella ciliata **L.**

PRICKLY ROELLA

Brown et al. (2003) showed that there was a significant increase in germination of the seeds of the South African plant, *R. ciliata*, which ranged from 32% for the control to 92% for seeds that were treated with aerosol smoke (see also Brown and Botha 2004).

Roella triflora **(Good) Adamson**

Roella triflora is native to South Africa. Treatment of its seeds with aerosol smoke significantly promoted germination in them (Brown et al. 2003; Brown and Botha 2004).

Wahlenbergia cernua **(Thunb.) A. DC.**

Wahlenbergia cernea is native to the Cape Floristic Region of South Africa. Aerosol smoke stimulated 45% of its seeds to germinate (Brown et al. 2003).

Wahlenbergia gracilenta **Lothian**

ANNUAL BLUEBELL

Germination of annual bluebell seed, collected from a *Eucalyptus baxteri* heathy-woodland in Victoria, Australia, significantly improved following a combined heat and smoke water treatment (Enright and Kintrup 2001).

Wahlenbergia gracilis **(G. Forst.) A. DC.**

SPRAWLING BLUEBELL

Sprawling bluebell is native to coastal and surrounding regions of central and eastern Australia. Germination doubled after the soil seed bank was exposed for 60 minutes to aerosol smoke (Read et al. 2000).

Wahlenbergia planiflora **P. J. Sm.**

Wahlenbergia planiflora is widely distributed in the woodland and forest communities of eastern Australia. Germination was inhibited by 54% when its seeds were treated with smoke water (10% dilution) (Clarke et al. 2000).

Wahlenbergia preissii **de Vriese**

Aerosol smoke treatments of broadcast seeds in rehabilitated mine sites in the southwest region of Western Australia had no effect on their germination (Roche et al. 1997a).

Wahlenbergia **sp.**

Soil samples from the Eden Burning Study Area, a dry sclerophyll forest in the Yalumba State Forest of New South Wales, Australia, were collected and air dried to test the effects of heat, smoke, and an interaction between the two cues on seeds from the seed bank. Samples exposed to heat treatment were incubated at 80°C for 60 minutes while those exposed to smoke were incubated in a room, where smoke was generated for 120 minutes. There was a marginally significant increase in germination as a result of the smoke treatment and only marginally significant inhibition due to an interaction between heat and smoke (Penman et al. 2008).

CAPRIFOLIACEAE

Weigela hortensis **C. A. Mey.**

GARDEN WEIGELA

The effects of aerosol smoke, heat, darkness, cold stratification, and combinations of smoke with each of the three other treatments on seed germination were examined in this study (Tsuyuzaki and Miyoshi 2009). Smoke was produced by burning Timothy hay (*Phleum pratense*), which was pumped through a 3.5 m cooling tube into a smoke chamber for approximately 5 minutes. The seeds were exposed to the smoke for 60 minutes. Those seeds exposed also to heat were incubated at 75°C for 25 minutes. The cold stratification process took 1 month, during which the seeds remained in an incubator set to 4°C. Where the dark treatment was concerned, the seeds were maintained in total darkness for the entire germination period. The aerosol smoke and dark treatments, when tested on their own, significantly inhibited seed germination, as did the combination of the smoke and heat treatment and the combination of the smoke and cold stratification treatment.

CARYOPHYLLACEAE

Dianthus **sp.**

PINKS

Seed germination in an unidentified species of *Dianthus*, which occurs in the Cape Floristic Region of South Africa, was not promoted by aerosol smoke (Brown et al. 2003).

Silene gallica

Silene cretica **L.**

CRETAN CATCHFLY

Silene cretica is native to the South African Cape Floristic Region. Germination was significantly promoted by aerosol smoke, but was low (Brown et al. 2003).

Silene gallica **L.**

COMMON CATCHFLY

Common catchfly is a global weed that occurs throughout much of western Europe, Russia, Georgia, India and western Asia. Germination was inhibited by powdered charred wood treatment (0.5 g per Petri dish) (Keeley and Keeley 1987).

Silene multinervia **S. Watson**

MANY NERVE CATCHFLY

Silene multinerva is a native of the Californian chaparrals of the United States. Keeley and Keeley (1987) reported that charate effectively promoted germination from 6% of seeds with no pretreatment to 44% when treated with charate. Germination of close to 100% was achieved by treating seeds with 5, 8, or 15 minutes of aerosol smoke. A 5% aqueous leachate of charred wood also increased germination, but was not as effective as the aerosol smoke treatment (Keeley and Fotheringham 1998a).

Silene regia **Sims**

ROYAL CATCHFLY

Germination of royal catchfly, native to the eastern and northcentral United States, was inhibited when its seeds were exposed to aerosol smoke for 60 minutes (Jefferson et al. 2007). Seventy percent germination was reported in the controls in comparison to 40% germination in treated seed.

Stellaria media **(L.) VIII.**

COMMON CHICKWEED

Common chickweed is a widespread temperate weed that most likely originated in Eurasia. Both karrikinolide (10^{-7} M) and smoke water (Capeseed, Cape Town, South Africa) significantly promoted germination in this species (Daws et al. 2007).

CASUARINACEAE

Allocasuarina fraseriana **(Miq.) L. A. Johnson.**

SHEOAK

Sheoaks occur in Western Australia's proteaceous heathland, jarrah forest and banksia woodland communities. Roche et al. (1997a) tested the effects of smoke water on sheoak seeds in disturbed soils of a bauxite mine site in jarrah forests and reported it had no effect on germination.

Allocasuarina littoralis **(Salisb.) L. A. Johnson.**

BLACK SHEOAK

This species occurs in the eastern states of Australia. Smoke water treatments of seed collected in the grassy woodlands of the New England Tableland of New South Wales had no effect on germination (Clarke et al. 2000).

CELASTRACEAE

Maytenus boaria **Mol.**

MAYTEN

Thirty minutes exposure to cool aerosol smoke significantly inhibited germination in this woody species of the Mediterranean matorral of central Chile (Gómez-González et al. 2008).

CISTACEAE

Cistus albidus **L.**

ROCK ROSE

The seeds of this rock rose were exposed to various temperatures, ranging from 80 to 150°C for periods of 5 and 10 minutes, and to two smoke water solutions of 1:1 and 1:10 concentrations, prepared according to Jager et al. (1996b). The smoke water solutions had no effect on germination percentage, rate, or seedling growth (Moreira et al. 2010). This species is native to southwestern Europe and northern Africa.

Cistus creticus **L.**

ROCK ROSE

Seeds of the rock rose, collected from a pine forest in the Bozburun peninsula of Muðla, Turkey, were treated to a combination of different heat shock treatments, ranging from 80 to 150°C, and smoke water treatments of 24 hours duration. The results of this study suggest that germination is promoted by heat shock between 100 and 150°C, but not by low temperatures and aqueous smoke (Tavsanoglu 2011).

Cistus crispus **L.**

SPOTTED ROCK ROSE

Spotted rock rose is commonly found growing in grasslands and open wood-lands of southwestern Europe and northwestern Africa. Pérez-Fernández and Rodríguez-Echeverría (2003) reported that germination was promoted when its seeds were exposed to aerosol smoke for 20 minutes. Nearly 40% of seed germi-nated when treated with smoke compared with less than 5% for the control group.

Cistus ladanifer **L.**

CRIMSON-SPOT ROCK ROSE

This species, like *C. crispus*, also occurs in grasslands and open woodlands of southwestern Europe and northwestern Africa. Germination of this species was increased to approximately 80% compared to approximately 15% for the control group. The seeds were treated with aerosol smoke for 20 minutes prior to sowing (Pérez-Fernández and Rodríguez-Echeverría 2003).

Cistus monspeliensis **L.**

MONTPELIER CISTUS

Cistus monspeliensis commonly occurs in the grasslands and open woodlands of southwestern Europe and northwestern Africa. Pérez-Fernández and Rodríguez-Echeverría (2003) reported that germination was promoted when its seeds were exposed to aerosol smoke for 20 minutes. In another study, the seeds were incu-bated for 24 hours in two smoke water solutions of 1:1 and 1:10 concentrations, pre-pared according to Jager et al. (1996b). These smoke water solutions had no effect on germination percentage or rate and did not affect seedling growth (Moreira et al. 2010). This species is native to northern Africa and southern Europe.

Cistus salviifolius **L.**

SALVIA CISTUS

Like the *Cistus* species above, *Salvia cistus* grows in the grasslands and open woodlands of southwestern Europe and northwestern Africa. Exposure to 10 or 20 minutes of aerosol smoke significantly promoted germination in this species (Pérez-Fernández and Rodríguez-Echeverría 2003). This species also occurs in the

Marmaris region of southwestern Turkey. Çatav et al. (2012) tested aqueous smoke solutions (24 hours of exposure) prepared from a variety of plant species on seeds gathered from that region, but did not, in contrast, observe any significant increases to seed germination or germination rate.

Fumana ericoides **(Cav) Gandg.**

FUMANA

The seeds of this species, collected from the Mediterranean Basin, were incubated for 24 hours in two smoke water solutions of 1:1 and 1:10 concentrations, prepared according to Jager et al. (1996b). The smoke water solutions had no effect on germination or seedling growth (Moreira et al. 2010).

Fumana thymifolia **Spach**

THYME ROCKROSE

Like *F. ericoides* above, the seeds of this species were collected from the Mediterranean Basin and incubated for 24 hours in two smoke water solutions of 1:1 and 1:10 concentrations to test the effects of the smoke water on germination. Neither of the two smoke water solutions had any effect on germination or seedling growth (Moreira et al. 2010).

Helianthemum syriacum **(Jacq.) Dum. Cours.**

COMMON ROCKROSE

Like *Fumana ericoides* and *F. thymifolia* above, the seeds of this species were collected from the Mediterranean Basin and also incubated for 24 hours in two smoke water solutions of 1:1 and 1:10 concentrations to test the effects of the smoke water on germination (Moreira et al. 2010). Neither of the smoke water solutions had any effect on germination percentage or rate and did affect seedling growth.

Lechea deckertii **Small**

PINWEED

Exposure to aerosol smoke for greater than 5 minutes inhibited seed germination in this upland Florida plant species (Lindon and Menges 2008).

Xolantha tuberaria **(L.) Gallego, Mŭnoz Garm. & C. Navarro**

SPOTTED ROCKROSE

The seeds of this species, collected from the Mediterranean Basin, were incubated for 24 hours in two smoke water solutions of 1:1 and 1:10 concentrations (as according to Jager et al. 1996b). Neither of the smoke water solutions had any effect on germination or seedling growth (Moreira et al. 2010). The smoke water solutions had no effect on germination or seedling growth (Moreira et al. 2010).

CLUSIACEAE

Hypericum gramineum **G. Forst.**

GRASSY ST. JOHNSWORT

Hypericum gramineum is native to Australia. Germination was promoted by exposing its seeds for 60 minutes to aerosol smoke (Read et al. 2000). Tang et al. (2003) reported in contrast, that 60-minute aerosol smoke treatments of soil samples containing the seeds of this species, collected from the Lamington National Park of Queensland, Australia, did not germinate in response to the smoke (Tang et al. 2003).

Hypericum **sp.**

Soil samples from the Eden Burning Study Area, a dry sclerophyll forest in the Yalumba State Forest of New South Wales, Australia, were collected and air dried to test the effects of heat, smoke, and an interaction between the two cues on seeds from the seed bank. Samples exposed to heat treatment were incubated at 80°C for 60 minutes while those exposed to smoke were incubated in a room, where smoke was generated for 120 minutes. None of the treatments had any effect on the germination of the seeds of an unidentified species of *Hypericum* (Penman et al. 2008).

COMBRETACEAE

Anogeissus leiocarpus **(DC.) Guill. & Perr.**

AXLE-WOOD TREE

Dayamba et al. (2008) reported that a combination of aerosol smoke treatment and heat shock significantly inhibited seed germination in this fire-sensitive species, which commonly occurs in the savannas of tropical Africa.

Combretum glutinosum **Perr. ex DC.**

Combretum glutinosum occurs all throughout West Africa. A combination of aerosol smoke and heat shock significantly improved seed germination in this fire-tolerant species (Dayamba et al. 2008).

Combretum nigricans **Leprieur ex Guill. & Perr.**

The exact distribution of *C. nigricans* is not known, but does occur in West Africa, where it is prized as a medicinal plant. Dayamba et al. (2008) reported that a combination of aerosol smoke treatment and heat shock significantly improved seed germination in this fire-tolerant species.

Pteleopsis suberosa **Engl. & Diels**

Dayamba et al. (2008) reported that seed germination in this fire-adapted species and commonly used medicinal plant in parts of Africa was significantly improved with aerosol smoke treatments.

Terminalia avicennioides **Guill. & Perr.**

This fire-adapted species is a native of Africa. Dayamba et al. (2008) reported that seed germination of its seeds was significantly improved with aerosol smoke treatments.

COMMELINACEAE

Commelina communis **L.**

ASIATIC DAYFLOWER

The effects of aerosol smoke, heat, darkness, cold stratification, and combinations of smoke with each of the three other treatments on seed germination were examined in this study (Tsuyuzaki and Miyoshi 2009). Smoke was produced by burning Timothy hay (*Phleum pratense*), which was pumped through a 3.5 m cooling tube into a smoke chamber for approximately 5 minutes. The seeds were exposed to the smoke for 60 minutes. Those seeds exposed also to heat were incubated at 75°C for 25 minutes. The cold stratification process took 1 month, during which the seeds remained in an incubator set to 4°C. Where the dark treatment was concerned, the seeds were maintained in total darkness for the entire germination period. All of control seeds failed to germinate. However, 90% germination occurred following the cold stratification treatment. None of the other treatments had any effect on germination. The seeds of this species display physiological dormancy.

CONVOLVULACEAE

Dichondra repens **J. R. Forst. & G. Forst.**

KIDNEY WEED

Kidney weed occurs throughout most of Australia. Aerosol smoke treatments of 60 minutes duration applied to soil seed banks in an open-cut coal mine of the Hunter Valley in New South Wales, Australia, did not significantly improve or promote seed germination (Read et al. 2000). In another study, soil samples from the Eden Burning Study Area, a dry sclerophyll forest in the Yalumba State Forest of New South Wales, Australia, were collected and air dried to test the effects of heat, smoke, and an interaction between the two cues on seeds from the seed bank (Penman et al. 2008). Samples exposed to heat treatment were incubated at 80°C for 60 minutes while those exposed to smoke were incubated in a room, where smoke was generated for 120 minutes. Only the heat treatment significantly improved germination of the seeds of this species. Neither the smoke treatment nor the interaction between the two cues had any effect.

CRASSULACEAE

Crassula capensis **Baill.**

Germination of *C. capensis*, which is native to the South African Cape Floristic Region, was enhanced by aerosol smoke treatment (Brown et al. 2003).

Crassula closiana **(Gay) Reiche**

STALKED CRASSULA

Stalked crassula seed, collected from a *Eucalyptus baxteri* heathy-woodland in Victoria, Australia, significantly improved following a treatment that combined heat, smoke water, and charred wood (Enright and Kintrup 2001).

Crassula coccinea **L.**

RED CRASSULA

This species occurs in South Africa's Cape Floristic Region. An application of aerosol smoke did not, however, improve seed germination (Brown et al. 2003).

CUPRESSACEAE

Actinostrobus acuminatus **(Parl.) F. Muell**

DWARF CYPRESS

Dwarf cypress is native to Western Australia's proteaceous heathland, banksia woodland, and jarrah forest communities. Dixon et al. (1995) reported that the germination of this species was significantly higher when the soil into which it was sown was exposed to 90 minutes of aerosol smoke.

Callitris intratropica **R. T. Baker & H. G. Sm.**

NORTHERN CYPRESS

Northern cypress is native to the northern regions of Western Australia, Northern Territory and Queensland, Australia. Roche et al. (1997b) showed that when fresh seeds were exposed to aerosol smoke for 60 minutes, applied after the seeds had been sown, they germinated more readily. Germination increased from 1.3% in the control to 13.9% when treated with smoke.

Juniperus procera **Hochst. ex Endl.**

AFRICAN PENCIL CEDAR

Smoke treatments were tested for their effects on overcoming photo-dormancy in the seeds of this species, but appeared ineffective (Tigabu et al. 2007). *Juniperus procera* is a native of the mountains of eastern Africa.

Widdringtonia cupressoides **(L.) Endl.**

MELANJE CEDAR

Seed of the South African fynbos species, *W. cupressoides*, was stimulated to germinate when the soil seed bank in which it occurred was exposed for 60 minutes to aerosol smoke (Brown et al. 1995).

Widdringtonia nodiflora **(L.) E. Powrie**

MOUNTAIN CYPRESS

Unlike *W. cupressoides*, germination in this species from the Cape Floristic Region of South Africa was not significantly promoted when the soil seed bank was exposed to aerosol smoke (Brown et al. 2003).

DILLENIACEAE

Hibbertia amplexicaulis **Steud.**

BUTTERCUP

Roche et al. (1997a) investigated the effects of smoke water and aerosol smoke application on the in situ germination of *H. amplexicaulis*—a native of the southwestern region of Western Australia, in particular its proteaceous heathlands, banksia woodlands, and jarrah forests. Both smoke water and aerosol smoke applications increased germination in the soil seed bank almost two-fold in comparison to the control. The ex situ germination of buttercup was also significantly improved by smoke (Dixon et al. 1995). Roche et al. (1997b) achieved the highest germination results when seeds that had been stored for 1 year prior to treatment were then exposed to aerosol smoke.

Hibbertia commutata **Steud.**

GUINEA FLOWER

Guinea flower is native to Western Australia's proteaceous heathland, banksia woodland, and jarrah forest regions. Seed germination significantly improved when the seeds had been stored in soil for 12 months, after which they are treated with aerosol smoke for 60 minutes (Roche et al. 1997b; Tieu et al. 2001b). Norman et al. (2006) reported, in contrast, that neither aerosol smoke nor smoke water treatments significantly promoted germination in this species. Roche et al. (1997a) reported that smoke water had no effect.

Hibbertia empetrifolia **(DC.) Hoogland**

TRAILING GUINEA FLOWER

Soil samples from the Eden Burning Study Area, a dry sclerophyll forest in the Yalumba State Forest of New South Wales, Australia, were collected and air dried to test the effects of heat, smoke, and an interaction between the two cues on seeds from the seed bank. Samples exposed to heat treatment were incubated at 80°C for 60 minutes while those exposed to smoke were incubated in a room, where smoke was generated for 120 minutes. Heat treatment of the seeds of this species significantly improved germination. The increase due to the smoke treatment was only marginally significant (Penman et al. 2008). An interaction between the two cues had no effect.

Hibbertia lasiopus **Benth.**

LARGE HIBBERTIA

Large hibbertia occurs in the jarrah forests of the southwestern region of Western Australia. Ex situ germination significantly improved when its seeds were sown in soil and then treated with aerosol smoke for 90 minutes (Dixon et al. 1995). Roche et al. (1997b) achieved a mean final germination of 29% after the seeds had been stored in the soil for 12 months and prior to aerosol smoke treatment of 60 minutes duration.

Hibbertia mylnei **Benth.**

Dixon et al. (1995) reported that 90 minutes of cold smoke treatment had no effect on seed germination of this Western Australian plant species.

Hibbertia obtusifolia **DC.**

GREY GUINEA FLOWER

This species has a wide distribution in the eastern states of Australia. Smoke water treatments of seeds collected from the grassy woodlands of the New England Tableland of New South Wales had no effect on germination (Clarke et al. 2000).

Hibbertia ovata **Steud.**

BUTTERCUP

Hibbertia ovata is native to the banksia woodlands and jarrah forests of the Southwest Botanical Province of Western Australia. Seed germination was significantly promoted after its seeds were exposed for 60 minutes to aerosol smoke (Roche et al. 1997b).

Hibbertia quadricolor **Domin.**

Hibbertia quadricolor is native to the jarrah forest and banksia woodland communities of the southwestern region of Western Australia. In 1995, Dixon and his colleagues discovered that smoke promoted germination in this species. Roche et al. (1997b) has since discovered that better germination will occur if the seeds are stored in the soil for 12 months prior to aerosol smoke treatment. Smoke water had no effect (Roche et al. 1997a).

Hibbertia racemosa **(Endl.) Gilg**

STALKED GUINEA FLOWER

Stalked guinea flower is a Western Australian plant that occurs in coastal areas from Exmouth to Esperance. Smoke water sprays, with concentrations of 50 and 100 mL/m², did not significantly promote germination of its seeds in an intact banksia woodland 20 km south of Perth, Western Australia (Lloyd et al. 2000).

Hibbertia riparia **Hoogland**

ERECT GUINEA FLOWER

Hibbertia riparia is commonly found growing in the coastal ranges of southeastern Australia. Storing seeds in soil for 12 months, and then exposing them to aerosol smoke for 60 minutes, significantly promoted germination (Roche et al. 1997b).

Hibbertia sericea **Benth.**

SILKY GUINEA FLOWER

Like *H. riparia* above, this species is also common in the coastal ranges of south-eastern Australia. Storing seeds in soil for 12 months, and then exposing them to aerosol smoke for 60 minutes, was enough to significantly promote germination (Roche et al. 1997b).

Hibbertia **sp.**

This rare and endangered *Hibbertia* species is native to the jarrah forests of southwestern Western Australia. Removal of the seed coat, followed by the addition of gibberellic acid to the growth medium (25 mg/L) and soaking the seeds in smoke water for 24 hours significantly promoted germination (Cochrane et al. 2002).

DIPSACACEAE

Scabiosa africana **L.**

CAPE SCABIOUS

This species occurs in South Africa's Cape Floristic Region. Treatment with aerosol smoke did not improve seed germination for this species (Brown et al. 2003).

DROSERACEAE

Drosera erythrorhiza **Lindl.**

RED INK SUNDEW

Red ink sundew is common to the jarrah forests of Western Australia. Dixon et al. (1995) reported that 90 minutes of cold smoke treatment had no effect on seed germination.

Drosera glanduligera **Lehm.**

SCARLET SUNDEW

Scarlet or pimpernel sundew seed, collected from a *Eucalyptus baxteri* heathy-woodland in Victoria, Australia, significantly improved following a treatment that combined heat, smoke water, and charred wood (Enright and Kintrup 2001).

Drosera gigantea **Lindl.**

GIANT SUNDEW

Giant sundew is commonly found in moist environments (i.e., swampy areas, granite outcrops) in the Southwest Botanical Province of Western Australia. An aerosol smoke treatment of 60 minutes of sown seeds had an inhibitory effect on the germination of this species (Roche et al. 1997b).

Drosera macrantha **Endl.**

CLIMBING SUNDEW

Climbing sundew is also common to the jarrah forests of Western Australia. Dixon et al. (1995) reported that 90 minutes of cold smoke treatment had no effect on seed germination. Roche et al. (1997b) reported that 60 minute exposure to cold smoke also had no significant effect on germination.

Drosera pallida **Lindl.**

PALE RAINBOW

Drosera pallida is endemic to Western Australia and grows on heathland, sand-plains, and along the coast. Freshly collected seeds of this Australian plant species, when exposed to 60 minutes of cool aerosol smoke, did not significantly affect germination (Roche et al. 1997b).

Drosera rotundifolia **L.**

ROUND LEAVED SUNDEW

The effects of aerosol smoke, heat, darkness, cold stratification, and combinations of smoke with each of the three other treatments on seed germination were examined in this study (Tsuyuzaki and Miyoshi 2009). Smoke was produced by burning Timothy hay (*Phleum pratense*), which was pumped through a 3.5 m cooling tube into a smoke chamber for approximately 5 minutes. The seeds were exposed to the smoke for 60 minutes. Those seeds exposed also to heat were incubated at 75°C for 25 minutes. The cold stratification process took 1 month, during which the seeds remained in an incubator set to 4°C. Where the dark treatment was concerned, the seeds were maintained in total darkness for the entire germination period. Seed germination was almost negligible (0%–0.5%) for the control, but was 37% germination following the cold stratification treatment. None of the other treatments had any effect on germination.

Drosera **sp.**

Soil samples from the Eden Burning Study Area, a dry sclerophyll forest in the Yalumba State Forest of New South Wales, Australia, were collected and air dried to test the effects of heat, smoke, and an interaction between the two cues on seeds from the seed bank. Samples exposed to heat treatment were incubated at 80°C for 60 minutes while those exposed to smoke were incubated in a room, where smoke was generated for 120 minutes. None of the treatments had any effect on the germination of the seeds of an unidentified species of *Drosera* (Penman et al. 2008).

EPACRIDACEAE

Acrotriche patula **R. Br.**

PRICKLY GROUND-BERRY

Acrotriche patula is common in adjacent regions to the coastline of the Great Australian Bight, Australia. Seed germination occurs best when the seeds are stored in soil for 12 months and then treated with aerosol smoke for 60 minutes (Roche et al. 1997b).

Andersonia bifida **(L.) Watson**

Cochrane et al. (2002) were able to achieve a final germination of 7% by soaking the seeds of this rare and endangered native of southwestern Western Australia's jarrah forest in concentrated smoke water for 24 hours. The seeds did not germinate at all without pretreatment.

Andersonia involucrata **Sond.**

Andersonia involucrata occurs in moist areas of banksia woodlands, jarrah forests and proteaceous heathlands of southwestern Western Australia. Ex situ germination increased significantly in response to aerosol smoke treatment of sown seeds (Roche et al. 1997b). Earlier trials by Dixon et al. (1995), in which the seeds were treated with cold smoke for 90 minutes in a tent, revealed that smoke had no significant effect on seed germination.

Andersonia latiflora **(F. Muell.) Benth.**

Andersonia latiflora is restricted to the banksia woodlands of the Swan Coastal Plains and northern jarrah forests of southwestern Western Australia. Like *A. involucrata*, ex situ germination increased significantly in response to aerosol smoke treatment of sown seeds (Roche et al. 1997b). Norman et al. (2006) reported, in contrast, that neither aerosol smoke nor smoke water treatments significantly promoted germination of its seeds.

Andersonia lehmanniana **Sond.**

Andersonia lehmannia is a native of southwestern Western Australia's proteaceous heathland, banksia woodlands, and jarrah forest regions. Ex situ germination significantly improved when the soil seed bank was treated with aerosol smoke for a period of 90 minutes (Dixon et al. 1995). Aerosol smoke treatments of broadcast seeds in rehabilitated mine sites in the southwest of Western Australia had no effect on their germination, however (Roche et al. 1997a).

Astroloma ciliatum **(Lindl.) Druce**

CANDLE CRANBERRY

An aerosol smoke treatment of broadcast seeds in a rehabilitated mine site in southwestern Western Australia had no effect on the germination of this species (Roche et al. 1997a).

Astroloma microcalyx **Sond.**

This species of *Astroloma* occurs in banksia woodlands in and around Perth, Western Australia. Germination of its seeds was not significantly promoted when they were soaked for 24 hours in 10% smoke water (Clarke and French 2005).

Astroloma pallidum **R. Br.**

KICK BUSH

Dixon et al. (1995) reported that 90 minutes of cold smoke treatment had no effect on seed germination of this Western Australian species.

Conostephium pendulum **Benth.**

PEARL FLOWER

Dixon et al. (1995) reported that 90 minutes of cold smoke treatment had no effect on seed germination of this Western Australian species.

Leucopogon capitellatus **DC.**

FLUFFY FLOWERS

Leucopogon capitellatus is native to the southwestern region of Western Australia. Germination significantly improved in response to aerosol smoke treatments of 60 minutes duration (Roche et al. 1997b). Germination of 2–3-year-old seeds treated with cold smoke for 90 minutes in a smoke tent was not, however, significantly improved (Dixon et al. 1995).

Leucopogon conostephioides **DC.**

Leucopogon conostephioides is widely distributed throughout the Southwest Botanical Province of Western Australia. Tieu et al. 2001b reported 26% germination when the deeply dormant seeds were buried in soil for 450 days and then smoke treated for 60 minutes.

Leucopogon crassiflorus **(F. Muell) Benth.**

FLUFFY FLOWERS

Dixon et al. (1995) reported that 90 minutes of cold smoke treatment had no effect on seed germination of this Western Australian species.

Leucopogon ericoides **(Sm.) R. Br.**

PINK BEARD HEATH

The effects of both heat and smoke on the germination of soil-stored seeds of this species, which occurs in the southeastern pasts of Australia, were researched by Wills and Read (2002). Smoke had no effect on seed germination.

Leucopogon gracilis **R. Br.**

Germination of *L. gracilis* seed, collected from a *Eucalyptus baxteri* heathy-woodland in Victoria, Australia, was significantly promoted following a combined heat and smoke water treatment (Enright and Kintrup 2001).

Leucopogon hirsutus **Sond.**

Dixon et al. (1995) reported that 90 minutes of cold smoke treatment had no effect on seed germination of this Western Australian species.

Leucopogon muticus **R. Br.**

BLUNT BEARD-HEATH

Smoke water treatments of seed collected from the grassy woodlands of the New England Tableland of New South Wales, Australia, had no effect on its germination (Clarke et al. 2000).

Leucopogon nutans **E. Pritz.**

DROOPING LEUCOPOGON

Leucopogon nutans is a native of Western Australian jarrah forest and banksia woodland ecosystems. Drooping leucopogon germination increased significantly in response to a smoke water treatment (1 L/m^2) of the soil seed bank of a bauxite mine site undergoing rehabilitation (Roche et al. 1997a). In addition, exposure of the seeds to aerosol smoke for 60 minutes also resulted in a significant increase in germination (Roche et al. 1997a,b).

Leucopogon obtectus **Benth.**

HIDDEN BEARD HEATH

Dixon et al. (1995) reported that 90 minutes of cold smoke treatment had no effect on seed germination of this Western Australian species.

Leucopogon propinquus **R. Br.**

Leucopogon propinquus occurs in moist areas of banksia woodlands, jarrah forests and proteaceous heathlands of the southwestern region of Western Australia. Ex situ germination increased significantly in response to aerosol smoke treatment of sown seeds (Roche et al. 1997b). Norman et al. (2006) reported, in contrast, that neither aerosol smoke nor smoke water treatments had any significant effect on seed germination. Dixon et al. (1995) reported that 90 minutes of cold smoke treatment also had no effect on seed germination.

Leucopogon **sp.**

Lloyd et al. (2000) showed that germination of an unidentified species of *Leucopogon* was promoted by in situ application of concentrated smoke water (50 or 100 mL/m²) to the soil seed bank of an area of banksia woodland in Western Australia.

Leucopogon verticillatus **R. Br.**

TASSEL PLANT

This species occurs in the jarrah (*Eucalyptus marginata*) forests of Western Australia. Neither aerosol smoke nor smoke water treatments significantly promoted germination of its seeds (Roche et al. 1997a,b; Norman et al. 2006). Dixon et al. (1995) reported that 90 minutes of cold smoke treatment also had no effect on seed germination.

Monotoca scoparia **(Sm.) R. Br.**

PRICKLY BROOM HEATH

Soil samples from the Eden Burning Study Area, a dry sclerophyll forest in the Yalumba State Forest of New South Wales, Australia, were collected and air dried to test the effects of heat, smoke, and an interaction between the two cues on seeds from the seed bank. Samples exposed to heat treatment were incubated at 80°C for 60 minutes while those exposed to smoke were incubated in a room, where smoke was generated for 120 minutes. Heat treatment of the seeds of this species significantly improved germination. The increase due to the smoke treatment was only marginally significant (Penman et al. 2008). An interaction between the two cues had no effect.

Sprengelia monticola **(DC.) Druce**

ROCK SPRENGELIA

Seed of this species is commonly found in soil seed banks of Sydney, Australia. Aerosol smoke treatments of 5, 10, and 20 minutes duration, in combination with and without heat shock, did not significantly increase germination in this species (Thomas et al. 2003).

Styphelia pulchella **(Stschegl.) Druce.**

Dixon et al. (1995) reported that 90 minutes of cold smoke treatment had no effect on seed germination of this Western Australian species.

Styphelia tenuiflora **Benth.**

PIN HEATH

This species occurs in the jarrah (*Eucalyptus marginata*) forests of Western Australia. Neither aerosol smoke nor smoke water treatments significantly promoted germination of its seeds (Roche et al. 1997a,b; Norman et al. 2006). Dixon et al. (1995) reported that 90 minutes of cold smoke treatment also had no effect on seed germination.

Woollsia pungens **F. Muell.**

The seeds of this species, when treated with smoke, did not exhibit any significant increase in germination, but did respond positively to a combination of smoke and heat shock (Thomas et al. 2007).

ERICACEAE

Arctostaphylos glandulosa **Eastw.**

EASTWOOD'S MANZANITA

Arctostaphylos glandulosa is a dominant species in the coastal chaparral communities of Mexico, California, and Oregon. This species would not germinate unless seeds were subjected to charate treatment (0.25 g of powdered charred wood on filter paper). Germination was still only 5% following treatment (Keeley 1987).

Arctostaphylos patula **Greene**

GREENLEAF MANZANITA

Greenleaf manzanita is common throughout the western United States, including the Californian chaparral, the Great Basin and much of the Sierra Nevada. Under dark conditions, charate significantly stimulated seed germination (Keeley 1987).

Arctostaphylos pungens **Kunth**

MEXICAN MANZANITA

Germination of up to 30% was achieved when the seeds of *A. pungens*, known locally in the pine-oak forests of Durango, Mexico, as Mexican manzanita, were treated to a variety of methods for promoting germination (Jurado et al. 2011). The best results were achieved when the seeds were stored at cold temperatures and then heat shocked at 120°C, exposed to aerosol smoke for 5 minutes and watered with a dilute charcoal solution.

Bejaria racemosa **Vent.**

FLYWEED

Exposure to aerosol smoke for greater than 5 minutes inhibited seed germination in this upland Florida plant species (Lindon and Menges 2008).

Calluna vulgaris **(L.) Hull**

DORMANT HEATHER

Calluna vulgaris is a common component of heathland habitats of Europe. Germination increased significantly in response to seed treatment with smoke water from four vegetation sources. Gibberellin treatment (100 mg/L) acted synergistically with the smoke water to further increase germination (Thomas and Davies 2002).

Comarostaphylis diversifolia **(Parry) Greene**

SUMMER HOLLY

Summer holly is native to the Californian chaparral and coastal sage scrub communities of the United States. This species is considered rare and at risk (IUCN 2013). The best germination results (69%) occurred when the seeds were not given a pretreatment. Charate increased germination to 46% versus 12% when its seeds had been heat treated at 100°C for 5 minutes (Keeley 1987).

Dracophyllum secundum **R. Br.**

This species can be found growing in moist rocky areas of central and southeastern New South Wales, Australia. Germination was not affected by heat shock treatments (Thomas et al. 2007). Aerosol smoke reversed the effect of heat (50 or 75°C for 5 minutes).

Epacris apsleyensis **Crowden**

APSLEY HEATH

This endangered epacrid grows in the dry sclerophyll forests of the east coast of Tasmania, Australia. Gilmour et al. (2000) achieved 73% germination under dark conditions when treating the seeds initially with heat shock (90°C for 10 minutes) and then germinating the seeds in Petri dishes containing 5% smoke water. The germination percentage of the control was 4.3%. The IUCN (2013) classification for this species is locally and globally vulnerable.

Epacris coriacea **A. Cunn. ex DC.**

Seed of this species is commonly found in soil seed banks of Sydney, Australia. Aerosol smoke treatments of 5, 10, and 20 minutes duration, in combination with and without heat shock, did not significantly increase germination of this species (Thomas et al. 2003).

Epacris crassifolia **R. Br.**

Epacris crassifolia grows in sandy crevices within rocky sandstone ledges in New South Wales, Australia. The negative effect of a heat shock treatment of 75°C for 5 minutes on "population 1" seed germination was significantly improved when the seeds were exposed to aerosol smoke for 5 and 10 minutes (Thomas et al. 2007). Neither smoke nor heat shock treatments alone had any effect on the seeds of "population 2."

Epacris impressa **Labill.**

PINK HEATH

Pink heath occurs in heath and woodland communities of southeastern Australia. Enright and Kintrup (2001) observed improved germination when smoke water was applied to the soil seed bank. In another study, soil samples from the Eden Burning Study Area, a dry sclerophyll forest in the Yalumba State Forest of New South Wales, Australia, were collected and air dried to test the effects of heat, smoke, and an interaction between the two cues on seeds from the seed bank (Penman et al. 2008). Samples exposed to heat treatment were incubated at 80°C for 60 minutes while those exposed to smoke were incubated in a room, where smoke was generated for 120 minutes. Both heat and smoke significantly improved germination for this species while an interaction between the two cues had no effect.

Epacris lanuginosa **Labill.**

WOOLLY HEATH

Woolly heath is common in southeast Australia (i.e., Tasmania, Victoria, and New South Wales), especially in wet heathland environments. Germination of seed under dark conditions and using heat shock in combination with smoke water treatments (as described for *E. aspleyensis*) was 43% in comparison to 0.7% for the control group (Gilmour et al. 2000).

Epacris longifolia **Cav.**

FUCHSIA HEATH

Fuchsia heath occurs in the sclerophyll forests of New South Wales, Australia. Five and 10 minutes exposure to aerosol smoke significantly promoted germination in seeds treated with heat at 50 and 75°C for 5 minutes (Thomas et al. 2007).

Epacris microphylla **var.** rhombifolia **R. Br.**

This heath is widespread along the eastern coastal regions of Australia. Thomas et al. (2007) showed that a combination of 20 minutes aerosol smoke and heat shock at 75°C for 5 minutes promoted germination in this species.

Epacris muelleri **Sonder**

This species has a small distribution on moist sandstone in the Blue Mountains and Wollemi National Park in New South Wales, Australia. Specific combinations of heat and smoke promoted germination of *E. muelleri*. Smoke alone had no effect. (Thomas et al. 2007).

Epacris obtusifolia **Sm.**

BLUNT-LEAF HEATH

Blunt-leaf heath is common in wet heathland and open forests of southeastern coastal regions and adjacent ranges of Australia (i.e., Queensland, New South Wales, Victoria, and Tasmania). Roche et al. (1997a) showed that the germination of this species increased in response to aerosol smoke application. Sixty-five percent germination under dark conditions was achieved by using smoke water and heat shock treatments (as described for *E. aspleyensis*) in comparison to 3% for the control (Gilmour et al. 2000). In contrast, Thomas et al. (2003) reported that aerosol smoke treatments of 5, 10, and 20 minutes duration, in combination with and without heat shock, had no significant effect on the germination of this species.

Epacris paludosa **R. Br.**

SWAMP HEATH

As the common name suggests, this species grows in damp habitats along the coastal regions of New South Wales, Victoria, and Tasmania, Australia. Germination of swamp heath was improved by aerosol smoke after heat shock treatments of 50 and 75°C for 5 minutes (Thomas et al. 2007).

Epacris pulchella **Cav.**

WALLUM HEATH

Wallum heath occurs along the coastal regions of New South Wales and southern Queensland, Australia. A significant interaction effect between smoke and heat treatments promoted germination in this species (Thomas et al. 2007). Smoke promoted germination in response to either no heat or a heat shock treatment of 75°C of five minutes duration. Smoke had no effect on germination when the seeds were exposed to heat at 50°C for 5 minutes and had inhibited germination when the seeds were treated at 100°C for 5 minutes.

Epacris purpurascens **R. Br.**

PORT JACKSON HEATH

Port Jackson heath is native to the dry sclerophyll forests of the southern coastal regions and adjacent ranges of Australia. Gilmour et al. (2000) achieved 75% germination using heat shock and smoke water treatments (as described for *E. aspleyensis*).

Epacris stuartii **Stapf**

STUART'S HEATH

A single population of this endangered heathland shrub exists in the coastal region of southeastern Tasmania, Australia. Stuart's heath is in danger of extinction (IUCN 2013). A maximum germination of 42% occurred when its seeds were exposed to heat shock (90°C for 10 minutes), and were then germinated on smoke paper in darkness (Keith 1997). The smoke paper was prepared by exposing filter paper to aerosol smoke for 90 minutes.

Epacris tasmanica **W. M. Curtis**

TASMAN HEATH

Tasman heath is endemic to the dry heathland environments of Tasmania, Australia. Gilmour et al. (2000) studied the effects of light and dark germination conditions, smoke water (0%, 5%, 10%, and 100%), and heat shock (90°C for 10 minutes) combinations on seed germination. The best germination results (49%) were achieved using a combination of heat shock, 5% smoke water, and dark conditions.

Erica baccans **L.**

BERRY HEATH

Berry heath is a native of the Cape Floristic Region of South Africa. Germination was promoted by 411% by exposing the seeds to aerosol smoke for a period of 30 minutes (Brown et al. 2003; Brown and Botha 2004).

Erica brachialis **Salisb.**

Seed germination of this South African native was not promoted by a 30-minute aerosol smoke treatment (Brown et al. 1993; Brown et al. 2003).

Erica caffra **L.**

SCENTED HEATH

Scented heath is a native of the Cape Floristic Region of South Africa. Germination was promoted when the seeds of this species were exposed to aerosol smoke for a period of 30 minutes (Brown et al. 2003; Brown and Botha 2004).

Erica canaliculata **Andr.**

HAIRY GREY HEATHER

This species is also a native of the Cape Floristic Region of South Africa. Germination was promoted when the seeds of this species were exposed to aerosol smoke for a period of 30 minutes (Brown et al. 1993; Brown et al. 1995; Brown et al. 2003; Brown and Botha 2004).

Erica capensis **Salter**

MONTEREY BAY HEATHER

This species is also a native of the Cape Floristic Region of South Africa. Germination was promoted when the seeds of this species were exposed to aerosol smoke for a period of 30 minutes (Brown et al. 1993; Brown et al. 1995; Brown et al. 2003; Brown and Botha 2004).

Erica capitata **L.**

Germination of *Erica capitata*, a native of South Africa, was significantly increased in response to treating its seeds with aerosol smoke for a period of 60 minutes (Brown et al. 1995). Treating it with aerosol smoke for 30 minutes had no significant effect (Brown et al. 1993; Brown et al. 2003). This species is classified as rare, both globally and locally (IUCN 2013).

Erica cerinthoides **L.**

FIRE HEATH

The seeds of fire heath, a South African fynbos species, germinated following 60 minutes of aerosol smoke treatment (Brown et al. 1995), but did not respond when exposed to 30 minutes of smoke (Brown et al. 1993; Brown et al. 2003).

Erica clavisepala **Guthrie & Bolus**

This species is also a native of the Cape Floristic Region of South Africa. Germination was promoted when the seeds of this species were exposed to aerosol smoke for a period of 30 minutes (Brown et al. 1993; Brown et al. 1995; Brown et al. 2003; Brown and Botha 2004). The IUCN (2013) has classified this species as rare and at risk.

Erica cruenta **[Soland.]**

BLOOD RED HEATH

Seed germination of this South African plant species was not promoted by 30 minutes of aerosol smoke treatment (Brown et al. 1993; Brown et al. 2003).

Erica curvirostris **Salisb.**

This species is also a native of the Cape Floristic Region of South Africa. Germination was promoted when the seeds of this species were exposed to aerosol smoke for a period of 30 minutes (Brown et al. 1993; Brown et al. 1995; Brown et al. 2003; Brown and Botha 2004).

Erica deflexa **Sinclair.**

This species is also a native of the Cape Floristic Region of South Africa. Germination was promoted after its seeds were exposed to aerosol smoke for a period of 30 minutes (Brown et al. 1993; Brown et al. 1995; Brown et al. 2003; Brown and Botha 2004).

Erica diaphana **Spreng.**

HEATH

This species is also a native of the Cape Floristic Region of South Africa. Germination was promoted when the seeds of this species were exposed to aerosol smoke for a period of 30 minutes (Brown et al. 1993; Brown et al. 1995; Brown et al. 2003; Brown and Botha 2004).

Erica dilatata **Wendl.f. ex Benth.**

Seed germination in this native to South Africa was significantly increased following 60 minutes of aerosol smoke treatment (Brown et al. 1995). Treating it with aerosol smoke for 30 minutes produced no significant effect (Brown et al. 1993).

Erica discolor **Andrews**

This species is also a native of the Cape Floristic Region of South Africa. Germination was promoted when the seeds of this species were exposed to aerosol smoke for a period of 30 minutes (Brown et al. 1993; Brown et al. 1995; Brown et al. 2003; Brown and Botha 2004).

Erica ericoides **L.**

This species is also a native of the Cape Floristic Region of South Africa. Germination was promoted after its seeds were exposed to aerosol smoke for 30 minutes (Brown et al. 1993; Brown et al. 1995; Brown et al. 2003; Brown and Botha 2004).

Erica formosa **Thunb.**

WHITE HEATH

This species is also a native of the Cape Floristic Region of South Africa. Germination was promoted when the seeds of this species were exposed to aerosol smoke for a period of 30 minutes (Brown et al. 1993; Brown et al. 1995; Brown et al. 2003; Brown and Botha 2004).

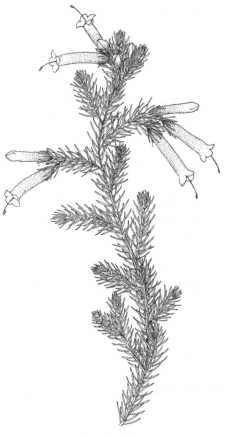

Erica discolor

Erica gallorum **L. Bolus**

Seed germination in this species was not promoted by 30 minutes of aerosol smoke treatment (Brown et al. 1993; Brown et al. 2003).

Erica glauca **var.** elegans **(Andr.) Bolus**

Seed germination in this native to South Africa was significantly promoted following 60 minutes of aerosol smoke treatment (Brown et al. 1995), while treating it with aerosol smoke for 30 minutes produced no significant effect (Brown et al. 1993; see also Brown et al. 2003).

Erica glauca **var.** glauca **Andrews**

CUP-AND-SAUCER HEATH

This species is also a native of the Cape Floristic Region of South Africa. Germination was promoted when its seeds were exposed to aerosol smoke for a period of 30 minutes (Brown et al. 1993; Brown et al. 1995; Brown et al. 2003; Brown and Botha 2004).

Erica glomiflora **Salisb.**

Aerosol smoke (30 minutes) and smoke water treatments of sown seeds significantly promoted the germination of this South African fynbos species (Brown 1993a; Brown et al. 1993; Brown et al. 1995; Brown et al. 2003; Brown and Botha 2004).

Erica grata **Guthrie & Bolus**

This species is also a native of the Cape Floristic Region of South Africa. Germination was promoted when the seeds of this species were exposed to aerosol smoke for a period of 30 minutes (Brown et al. 1993; Brown et al. 1995; Brown et al. 2003; Brown and Botha 2004).

Erica halicacaba **L.**

Seed germination of this South African plant species was not promoted by 30 minutes of aerosol smoke treatment (Brown et al. 1993; Brown et al. 2003).

Erica hebecalyx **Benth.**

Aerosol smoke (30 minutes) and smoke water treatments of sown seeds significantly promoted the germination of this South African fynbos species (Brown 1993; Brown et al. 1993; Brown et al. 1995; Brown et al. 2003; Brown and Botha 2004).

Erica hirtiflora **Curtis**

This species is also a native of the Cape Floristic Region of South Africa. Germination was promoted when the seeds of this species were exposed to aerosol smoke for a period of 30 minutes (Brown et al. 1993; Brown et al. 1995; Brown et al. 2003; Brown and Botha 2004).

Erica junonia **var.** minor **Bolus**

Seed germination of this native of the Cape Floristic Region of South Africa was significantly promoted following 60 minutes of aerosol smoke treatment (Brown et al. 1995). Treating it with aerosol smoke for 30 minutes produced no significant effect (Brown et al. 1993; Brown et al. 2003; Brown and Botha 2004). This species is classified as endangered at both the local and global scale (IUCN 2013).

Erica lateralis **Willd.**

BUTTON HEATH

This species is also a native of the Cape Floristic Region of South Africa. Germination was promoted when the seeds of this species were exposed to aerosol smoke for a period of 30 minutes (Brown et al. 1993; Brown et al. 1995; Brown et al. 2003; Brown and Botha 2004).

Erica latiflora **(L.) Bolus**

Germination of this South African fynbos species significantly increased when sown seeds were exposed to aerosol smoke for 30 minutes (Brown et al. 1993; Brown et al. 1995).

Erica leptopus **Benth.**

This species occurs in the Cape Floristic Region of South Africa. Aerosol smoke had no effect on its seed germination (Brown et al. 2003).

Erica longiflora **Salisb.**

Germination of this South African fynbos species significantly increased when sown seeds were exposed to aerosol smoke for 30 minutes (Brown et al. 1993; Brown et al. 1995).

Erica longifolia **Willd.**

LONG-LEAVED HEATH

Aerosol smoke (30 minutes) and smoke water treatments of sown seeds significantly promoted the germination of this South African fynbos species (Brown 1993; Brown et al. 1993; Brown et al. 1995; Brown et al. 2003; Brown and Botha 2004).

Erica multiflora **L.**

BREZO

The seeds of this species, collected from the Mediterranean Basin, were incubated for 24 hours in two smoke water solutions of 1:1 and 1:10 concentrations, prepared according to Jager et al. (1996b). The smoke water solutions significantly increased germination percentage and rate, but had no effect on seedling growth (Moreira et al. 2010).

Erica nudiflora **L.**

This species is also a native of the Cape Floristic Region of South Africa. Germination was promoted when its seeds were exposed to aerosol smoke for a period of 30 minutes (Brown et al. 1993; Brown et al. 1995; Brown et al. 2003; Brown and Botha 2004).

Erica oatesii **Rolfe**

TREE HEATH

This species is also a native of the Cape Floristic Region of South Africa. Germination was promoted after its seeds were for 30 minutes exposed to aerosol smoke (Brown et al. 1993; Brown et al. 1995; Brown et al. 2003; Brown and Botha 2004).

Erica oblongiflora **Benth.**

Seed germination in this fynbos species of the Cape Floristic Region of South Africa was not significantly promoted following 30 minutes of aerosol smoke treatment (Brown et al. 1993; Brown et al. 2003).

Erica patersonia **Thunb.**

MIELIE HEATH

Mielie heath occurs in the Cape Floristic Region of South Africa. Aerosol smoke had no effect on its seed germination (Brown et al. 2003).

Erica perlata **Sinclair.**

This species is also a native of the Cape Floristic Region of South Africa. Germination was promoted after its seeds were exposed to aerosol smoke for 30 minutes (Brown et al. 1993; Brown et al. 1995; Brown et al. 2003; Brown and Botha 2004).

Erica perspicua **J. C. Wendl.**

PRINCE OF WALES HEATH

Seed germination in this fynbos species of the Cape Floristic Region of South Africa was not significantly improved following 30 minutes of aerosol smoke treatment (Brown and Botha 2004).

Erica peziza **Lodd.**

KAPOKKIE HEATH

Seed germination in this fynbos species of the Cape Floristic Region of South Africa was not significantly promoted by 30 minutes of aerosol smoke treatment (Brown et al. 1993; Brown et al. 2003).

Erica phylicifolia **Salisb.**

AUTUMN HEATH

This species is also a native of the Cape Floristic Region of South Africa. Germination was promoted when the seeds of this species were exposed to aerosol smoke for a period of 30 minutes (Brown et al. 1993; Brown et al. 1995; Brown et al. 2003; Brown and Botha 2004).

Erica pillansii **Bolus**

Seed germination in this South African plant species was not promoted by 30 minutes of aerosol smoke treatment (Brown et al. 1993; Brown et al. 2003).

Erica pinea **Thunb.**

PINE-LEAVED ERICA

This species is also a native of the Cape Floristic Region of South Africa. Germination was promoted after its seeds were exposed to aerosol smoke for

30 minutes (Brown et al. 1993; Brown et al. 1995; Brown et al. 2003; Brown and Botha 2004).

Erica plukenettii **L.**

SNOTBEL

This species is also a native of the Cape Floristic Region of South Africa. Germination was promoted following an aerosol smoke treatment of 30 minutes duration (Brown et al. 1993; Brown et al. 1995; Brown et al. 2003; Brown and Botha 2004).

Erica porteri **Compton**

Seed germination in this fynbos species of the Cape Floristic Region of South Africa was not significantly improved after 30 minutes of aerosol smoke treatment (Brown and Botha 2004).

Erica recta **Bolus**

Germination of *E. recta* (native to South Africa) significantly increased following 60 minutes exposure to aerosol smoke (Brown et al. 1995), while 30 minutes of exposure had no effect (Brown et al. 1993; Brown et al. 2003).

Erica sessiliflora **L. f.**

GREEN HEATH

This species is also a native of the Cape Floristic Region of South Africa. Germination was significantly promoted following 30 minutes to aerosol smoke treatment (Brown et al. 1993; Brown et al. 1995; Brown et al. 2003; Brown and Botha 2004).

Erica simulans **Dulfer**

This species is also a native of the Cape Floristic Region of South Africa. Germination was promoted after its seeds were exposed to aerosol smoke for a period of 30 minutes (Brown et al. 1993; Brown et al. 1995; Brown et al. 2003; Brown and Botha 2004).

Erica sitiens **Klotzsch**

This species is also a native of the Cape Floristic Region of South Africa. Germination was promoted after its seeds were exposed to aerosol smoke for a period of 30 minutes (Brown et al. 1993; Brown et al. 1995; Brown et al. 2003; Brown and Botha 2004).

Erica spectabilis **Klotzsch ex Benth.**

This species is also a native of the Cape Floristic Region of South Africa. Germination was promoted following a 30 minute exposure to aerosol smoke (Brown et al. 1993; Brown et al. 1995; Brown et al. 2003; Brown and Botha 2004).

Erica sphaeroidea **Dulfer**

This species is also a native of the Cape Floristic Region of South Africa. Germination was promoted if its seeds were exposed to aerosol smoke for a period of 30 minutes (Brown et al. 1993; Brown et al. 1995; Brown et al. 2003; Brown and Botha 2004).

Erica taxifolia **Ait.**

YEW-LEAVED ERICA

This species is also a native of the Cape Floristic Region of South Africa. Germination was after its seeds were exposed to aerosol smoke for a period of 30 minutes (Brown et al. 1993; Brown et al. 1995; Brown et al. 2003; Brown and Botha 2004).

Erica terminalis **Salisb.**

CORSICAN HEATH

The seeds of this species were incubated for 24 hours in two smoke water solutions of 1:1 and 1:10 concentrations, prepared according to Jager et al. (1996b). The smoke water solutions significantly increased germination percentage and rate, but had no effect on seedling growth (Moreira et al. 2010). This species is native to southern Europe and northern Africa.

Erica thomae **L. Bolus**

This species is also a native of the Cape Floristic Region of South Africa. Germination was promoted after its seeds were exposed to aerosol smoke for a period of 30 minutes (Brown et al. 1993; Brown et al. 1995; Brown et al. 2003; Brown and Botha 2004).

Erica tumida **Ker-Gawl.**

TUMID HEATH

This species is also a native of the Cape Floristic Region of South Africa. Germination was promoted following a 30 minute exposure to aerosol smoke (Brown et al. 1993; Brown et al. 1995; Brown et al. 2003; Brown and Botha 2004).

Erica turgida **Salisb.**

Erica turgida is native to South Africa. Germination occurred in significantly greater numbers if seeds were initially treated with aerosol smoke (Brown et al. 1995). In contrast, Brown et al. (2003) reported that aerosol smoke had no effect on germination. The IUCN (2013) has classified this species as a taxon that is suspected of having recently become extinct.

Erica umbellata **L.**

DWARF SPANISH HEATH

The seeds of this species, collected from the Mediterranean Basin, were incubated for 24 hours in two smoke water solutions of 1:1 and 1:10 concentrations, prepared according to Jager et al. (1996b). The smoke water solutions significantly increased germination percentage and rate, but had no effect on seedling growth (Moreira et al. 2010). This species is native to southern Europe and northern Africa.

Erica verecunda **Salisb.**

This heath occurs in the Cape Floristic Region of South Africa. Aerosol smoke had no effect on its seed germination (Brown et al. 2003).

Erica versicolor **J. C. Wendl.**

OUTENIQUA HEATHER

This species is also a native of the Cape Floristic Region of South Africa. Germination was promoted after its seeds were exposed to aerosol smoke for a period of 30 minutes (Brown et al. 1993; Brown et al. 1995; Brown et al. 2003; Brown and Botha 2004).

Erica vestita **Thunb.**

TREMBLING HEATH

This species is also a native of the Cape Floristic Region of South Africa. Germination was promoted following a 30 minute exposure to aerosol smoke (Brown et al. 1993; Brown et al. 1995; Brown et al. 2003; Brown and Botha 2004).

Gaultheria miqueliana **Takeda**

MIQUEL'S SPICY WINTERGREEN

The effects of aerosol smoke, heat, darkness, cold stratification, and combinations of smoke with each of the three other treatments on seed germination were examined in this study (Tsuyuzaki and Miyoshi 2009). Smoke was produced by burning Timothy hay (*Phleum pratense*), which was pumped through a 3.5 m cooling tube into a smoke chamber for approximately 5 minutes. The seeds were exposed to the smoke for 60 minutes. Those seeds exposed also to heat were incubated at 75°C for 25 minutes. The cold stratification process took 1 month, during which the seeds remained in an incubator set to 4°C. Where the dark treatment was concerned, the seeds were maintained in total darkness for the entire germination period. The heat (70%) and dark (0%) treatments both significantly inhibited germination, as did the combined smoke and cold stratification (39%) treatment. Germination in the control group was 83%.

Leucothoe grayana **Maxim. var.** oblongifolia **(Miq.) Ohwi**

LEUCOTHOE

The effects of aerosol smoke, heat, darkness, cold stratification, and combinations of smoke with each of the three other treatments on seed germination were examined in this study (Tsuyuzaki and Miyoshi 2009; for details about the tests performed, see *Gaultheria miqueliana* above). Smoke, heat, cold stratification, and the combined smoke and heat treatments resulted in higher germination percentages of 87%, 84%, 83%, and 75%, respectively while the dark treated seeds mostly failed to germinate (0%–0.5%). This was compared to 67% for the control group.

Lissanthe strigosa (Sm.) R. Br. ssp. subulta (R. Br.) J. M. Powell

PEACH HEATH

Smoke water treatments of seeds collected from the grassy woodlands of the New England Tableland of New South Wales, Australia, had no significant effect on germination (Clarke et al. 2000).

Melichrus urceolatus R. Br.

URN HEATH

Smoke water treatments of seed collected from the grassy woodlands of the New England Tableland of New South Wales, Australia, had no significant effect on germination (Clarke et al. 2000).

Tripetaleia paniculata Siebold & Zucc.

TRIPETALEIA

The effects of aerosol smoke, heat, darkness, cold stratification, and combinations of smoke with each of the three other treatments on seed germination were examined in this study (Tsuyuzaki and Miyoshi 2009; for details about the tests performed, see *Gaultheria miqueliana* above). There was no significant germination following any of the treatments except for cold stratification, in which 81% of the seeds germinated. This species is native to Japan.

EUPHORBIACEAE

Adriana quadripartita (Labill.) Muell. Arg.

BITTER BUSH

Dixon et al. (1995) reported that 90 minutes of cold smoke treatment had no effect on seed germination on this Western Australian species.

Adriana tomentosa Gaudich.

Dixon et al. (1995) reported that 90 minutes of cold smoke treatment had no effect on seed germination on this Western Australian species.

Amperea xiphoclada (Spreng.) Druce

BROOM SPURGE

Soil samples from the Eden Burning Study Area, a dry sclerophyll forest in the Yalumba State Forest of New South Wales, Australia, were collected and air dried to test the effects of heat, smoke, and an interaction between the two cues on seeds from the seed bank. Samples exposed to heat treatment were incubated at 80°C for

60 minutes while those exposed to smoke were incubated in a room, where smoke was generated for 120 minutes. Only the smoke treatment significantly improved the germination of the seeds of this species (Penman et al. 2008). Neither heat nor the interaction between the two cues had any effect.

Glochidion ferdinandi (Müll. Arg.) F. M. Bailey

CHEESE TREE

The cheese tree is native to eastern Australia. A 60 minute aerosol smoke treatment of soil samples containing seeds of this species, collected across forest edges between subtropical rainforests and eucalypt forests in the Lamington National Park of Queensland, Australia, did not significantly promote germination (Tang et al. 2003).

Mercurialis annua L.

ANNUAL MERCURY

This native of Europe is a weed in parts of Australia. A commercially available smoke water solution "Seed Starter" did not significantly improve seed germination (Adkins and Peters 2001).

Phyllanthus calycinus Labill.

FALSE BORONIA

False boronia occurs on sandy soils, mostly in the coastal heathland, jarrah forest, and banksia woodland communities of the southwestern region of Western Australia. Germination was significantly improved in response to in situ aerosol smoke treatments of the soil surface (Roche et al. 1997a). A smoke water treatment to the soil seed bank of this species also improved germination on a bauxite mine site undergoing rehabilitation (Roche et al. 1997a). Interestingly, Norman et al. (2006) reported that neither aerosol smoke nor smoke water treatments significantly promoted germination in false boronia seeds.

Phyllanthus tenellus Roxb.

LONG-STALK LEAF-FLOWER

This species is thought to be native to Madagascar, but occurs as a weed in several countries, including Australia and the United States. A 60 minute aerosol smoke treatment of soil samples containing seeds of this species, collected across forest edges between subtropical rainforests and eucalypt forests in the Lamington National Park of Queensland, Australia, did not germinate (Tang et al. 2003).

Phyllanthus virgatus G. Forst.

This species is common throughout Australia and parts of Asia and Melanesia. Smoke water treatments of seeds collected from the grassy woodlands of the New

England Tableland of New South Wales, Australia, had no effect on its germination (Clarke et al. 2000).

Poranthera huegelii **Kltzsch**

Aerosol smoke treatments of broadcast seeds in rehabilitated mine sites in the southwest of Western Australia had no effect on their germination (Roche et al. 1997a).

Poranthera microphylla **Brongn.**

SMALL PORANTHERA

Poranthera microphylla is widely distributed throughout the mediterranean, arid and tropical regions of Western Australia, and most likely the adjacent states of the Northern Territory and South Australia. The germination of this species increased in response to in situ aerosol smoke applications to the soil surface in the jarrah forest in southwestern Western Australia (Roche et al. 1997a). In another study, soil samples from the Eden Burning Study Area, a dry sclerophyll forest in the Yalumba State Forest of New South Wales, Australia, were collected and air dried to test the effects of heat, smoke, and an interaction between the two cues on seeds from the seed bank (Penman et al. 2008). Samples exposed to heat treatment were incubated at 80°C for 60 minutes while those exposed to smoke were incubated in a room, where smoke was generated for 120 minutes. Both heat and smoke treatments on their own significantly improved germination for the seeds of this species while an interaction between the two cues had no effect.

Ricinocarpus glaucus **Endl.**

WEDDING BUSH

Distribution information was not found, but this species is thought to be a coastal species of Western Australia. Germination of its seeds occurred only after they were treated with aerosol smoke for a period of 60 minutes (Roche et al. 1997b).

FABACEAE

Abrus precatorius **L.**

ROSARYPEA

Exposure to aerosol smoke for 1 and 5 minutes significantly promoted seed germination in this upland Florida plant species (Lindon and Menges 2008).

Acacia angustissima **(Mill.) Kuntze var.** angustissima

PRAIRIE ACACIA

A variety of different fire cues were tested on the seeds of this species, which were collected from a mixed forest located in a mountainous subtropical area of Mexico (Zuloaga-Aguilar et al. 2011). These included heat shock (100:15 sand: water [w:w] substrate at 120°C for 5 minutes), soaking the seeds in smoke water for 3 hours (prepared by burning 150 g of mixed forest litter and bubbling the resultant smoke into 1.5 L of distilled water, and adjusting the pH to 5 using sodium hydroxide) or ash (1.5 g of fine ash was added to the agar plates used to germinate the seeds). Combinations of these treatments were also tested for their effects on seed germination. The results revealed that heat shock, when applied on its own, significantly promoted germination, while the smoke water and ash treatments had no significant effect. Combinations of heat shock with smoke water and/or ash treatments all significantly improved germination.

Acacia catechu **(L. f.) Willd.**

CUTCHTREE

In this study, the seeds of cutchtree were collected from a dry deciduous forest in India and later exposed to aerosol smoke until they turned completely brown. The seeds were arranged onto filter papers that were hung in fumigation tents, in which litter collected from locally grown dry deciduous trees was burned to generate the smoke. Germination percentages and the germination velocity index of seeds were measured (Singh and Raizada 2010). The results of this study revealed that germination percentage was significantly increased following smoke treatment, with 95% germination compared to 60% for the control group. The germination velocity index was not significantly affected by the aerosol smoke.

Acacia caven **(Mol.) Mol.**

ESPINILLO (SPANISH)

Thirty minutes exposure to cool aerosol smoke significantly promoted germination in this woody species of the Mediterranean matorral of central Chile (Gómez-González et al. 2008).

Acacia cyclops **A. Cunn ex Don**

COASTAL WATTLE

Coastal wattle, as the name suggests, occurs along the coast and adjacent regions of the southwestern region of Western Australia and Southern Australia. Ex situ aerosol smoke applications significantly increased germination in this species (Roche et al. 1997b).

Acacia dealbata **Link**

SILVER WATTLE

This wattle is common throughout Australia. Smoke water treatments of seeds collected from the grassy woodlands of the New England Tableland of New South Wales, Australia, had no effect on its germination (Clarke et al. 2000).

Acacia hebeclada **DC.**

CANDLE THORN

Germination in seeds treated with smoke-derived karrikinolide (10^{-7} M) at different light conditions was significantly promoted in this deciduous shrub (Kulkarni et al. 2007a). Incubating the seeds with karrikinolide for 10 days in constant dark conditions and at 25 ±0.5°C resulted in a greater vigor index and seedling mass.

Acacia huegelii **Benth.**

This species occurs in banksia woodlands in Western Australia. Germination of its seeds was not significantly promoted when they were soaked in 10% smoke water for 24 hours (Clarke and French 2005).

Acacia longifolia **(Andrews) Willd.**

LONG LEAVED WATTLE

Soil samples from the Eden Burning Study Area, a dry sclerophyll forest in the Yalumba State Forest of New South Wales, Australia, were collected and air dried to test the effects of heat, smoke, and an interaction between the two cues on seeds from the seed bank. Samples exposed to heat treatment were incubated at 80°C for 60 minutes while those exposed to smoke were incubated in a room, where smoke was generated for 120 minutes. Only the heat treatment significantly improved the germination of the seeds of this species (Penman et al. 2008). Neither smoke nor the interaction between the two cues had any effect.

Acacia mearnsii **De Wild.**

BLACK WATTLE

Seeds of this fast-growing leguminous Australian native, when exposed to constant darkness and treated with smoke-derived karrikinolide (10^{-7} M), exhibited greater germination than the control group (Kulkarni et al. 2007a). Incubating the seeds with karrikinolide (10^{-7} M) for 10 days in constant dark conditions and at 25 ±0.5°C also resulted in a greater vigor index and seedling mass.

Acacia myrtifolia **(Sm.) Willd.**

MYRTLE WATTLE

Soil samples from the Eden Burning Study Area, a dry sclerophyll forest in the Yalumba State Forest of New South Wales, Australia, were collected and air dried to test the effects of heat, smoke, and an interaction between the two cues on seeds from the seed bank. Samples exposed to heat treatment were incubated at 80°C for 60 minutes while those exposed to smoke were incubated in a room, where smoke was generated for 120 minutes. All three treatments significantly improved germination of the seeds of this species (Penman et al. 2008).

Acacia robusta **Burch.**

STINK BARK ACACIA

Seeds incubated with karrikinolide (10^{-7} M) for 10 days in constant darkness and at 25 ±0.5°C exhibited greater vigor index and seedling mass (Kulkarni et al. 2007a).

Acacia **sp.**

WATTLE

Germination of an unidentified species of wattle increased significantly in response to aerosol smoke treatments of the seeds (Roche et al. 1997a). A 60 minute aerosol smoke treatment of soil samples containing seeds of another unidentified wattle, collected across forest edges between subtropical rainforests and eucalypt forests in the Lamington National Park of Queensland, Australia, did not germinate (Tang et al. 2003). These two wattles are most likely two different species.

Acacia terminalis **(Salisb.) J. F. Macbr.**

SUNSHINE WATTLE

Soil samples from the Eden Burning Study Area, a dry sclerophyll forest in the Yalumba State Forest of New South Wales, Australia, were collected and air dried to test the effects of heat, smoke, and an interaction between the two cues on seeds from the seed bank. Samples exposed to heat treatment were incubated at 80°C for 60 minutes while those exposed to smoke were incubated in a room, where smoke was generated for 120 minutes. Only the heat treatment significantly improved the germination of the seeds of this species (Penman et al. 2008). Neither smoke nor the interaction between the two cues had any effect.

Amorpha canescens Pursh

LEAD PLANT

Lead plant seed, collected from a tallgrass prairie in the Midwest of the United States, did not germinate in response to aerosol smoke treatments of 1, 10, or 60 minutes duration (Jefferson et al. 2007).

Anthyllis cytisoides **L.**

ALBAIDA

Seeds were incubated for 24 hours in two smoke water solutions of 1:1 and 1:10 concentrations, prepared according to Jager et al. (1996b). The smoke water solutions had no effect on germination or seedling growth (Moreira et al. 2010). This species is common in southern Europe.

Anthyllis lagascana **Benedí**

ALBAIDA ROSA

Like A. *cytisoides* above, the seeds of this species were incubated for 24 hours in two smoke water solutions of 1:1 and 1:10 concentrations, prepared according to Jager et al. (1996b). The smoke water solutions had no effect on germination percentage, rate, or seedling growth (Moreira et al. 2010). This species, which occurs in parts of Africa and the Iberian Peninsula, is considered a threatened species.

Aotus ericoides (Vent.) G. Don.

COMMON AOTUS

Soil samples from the Eden Burning Study Area, a dry sclerophyll forest in the Yalumba State Forest of New South Wales, Australia, were collected and air dried to test the effects of heat, smoke, and an interaction between the two cues on seeds from the seed bank. Samples exposed to heat treatment were incubated at 80°C for 60 minutes while those exposed to smoke were incubated in a room, where smoke was generated for 120 minutes. Smoke treatment of the seeds of this species significantly improved their germination. An increase due to the heat treatment was only marginally significant (Penman et al. 2008). An interaction between the two cues had no effect.

Astragalus crassicarpus Nutt.

GROUNDPLUM MILKVETCH

Chou et al. (2012) tested smoke water, heat and combinations of both heat and smoke on the germination percentage and mean germination times of this species. Smoke treatments were applied by soaking seeds for 20 hours in the commercially available Regen 2000® smoke water solution, at concentrations of 1:5, 1:10, or 1:100 (v/v). Seeds were exposed to heat treatments of 50 or 80°C for 5 minutes. Germination percentages were significantly promoted by 1:100 dilutions of smoke water. Mean germination times were significantly reduced when seeds were treated with the 1:10 dilution of smoke water. The seeds of this species did not respond to heat treatments, nor was there a significant interaction between heat and smoke treatments.

Baptisia australis (L.) R. Br.

BLUE WILD INDIGO

Seed of blue wild indigo, collected from a tallgrass prairie in the Midwest of the United States, did not germinate in response to aerosol smoke treatments of 1, 10, or 60 minutes duration (Jefferson et al. 2007).

Bauhinia variegata L.

ORCHID TREE

In this study, the seeds of *B. variegata* were collected from a dry deciduous forest in India and later exposed to aerosol smoke until they turned completely brown. The seeds were arranged onto filter papers that were hung in fumigation tents, in which litter collected from locally grown dry deciduous trees was burned to generate the smoke. Germination percentages and the germination velocity index of seeds were measured (Singh and Raizada 2010). The results revealed that both germination percentage (50%) and germination velocity index (1.79) significantly increased following the smoke treatment. This was compared to 35% and 1.08, respectively, for the control group.

Bossiaea aquifolium **Benth.**

WATER BUSH

The distribution of this species occurs in the proteaceous heathland, jarrah forest and banksia woodland communities of the southwest of Western Australia. Germination was significantly promoted when the soil seed bank was treated with smoke water (Roche et al. 1997a).

Bossiaea eriocarpa **Benth.**

COMMON BROWN PEA

Bossiaea eriocarpa is native to Western Australia's banksia woodland regions of the southwest. Smoke water sprays, with concentrations of 50 and 100 mL/m^2, did not significantly promote germination of seeds occurring naturally in an intact woodland 20 km south of Perth, Western Australia (Lloyd et al. 2000). Germination of its seeds was also unaffected by soaking them for 24 hours in 10% smoke water (Clarke and French 2005).

Bossiaea ornata **(Lindl.) Benth.**

A 1 L/m^2 application of undiluted smoke water to the soil surface of a rehabilitated bauxite mine in the southwest of Western Australia had no significant effect on the germination of the seeds in the soil seed bank (Roche et al. 1997a).

Calicotome villosa **(Poir.) Link.**

SPINY BROOM

Aqueous smoke solutions (24 hours exposure) prepared from a variety of plant species did not significantly increase seed germination or germination rate for this woody species from the Marmaris region of southwestern Turkey (Çatav et al. 2012).

Calliandra longipedicellata **(McVaugh) Macqueen and H. M. Hern**

CALLIANDRA

A variety of different fire cues were tested on the seeds of this species, which were collected from a mixed forest located in a mountainous subtropical area of Mexico (Zuloaga-Aguilar et al. 2011). These included heat shock (100:15 sand: water [w:w] substrate at 120°C for 5 minutes), soaking the seeds in smoke water for 3 hours (prepared by burning 150 g of mixed forest litter and bubbling the resultant smoke into 1.5 L of distilled water, and adjusting the pH to 5 using sodium hydroxide) or ash (1.5 g of fine ash was added to the agar plates used to germinate the seeds). Combinations of these treatments were also tested for their effects on seed germination. The results showed that heat shock, when applied on its own, as well as a combination of heat shock and ash treatments significantly improved germination.

Coronilla minima **L.**

CROWN VETCH

The seeds of crown vetch were incubated for 24 hours in two smoke water solutions of 1:1 and 1:10 concentrations, prepared according to Jager et al. (1996b). The smoke water solutions had no effect on germination percentage or rate and did not affect seedling growth (Moreira et al. 2010). This species is common in the Mediterranean Basin.

Crotalaria longirostrata **Hook. & Arn.**

CHIPILÍN

Like *Calliandra longipedicellata* above, a variety of different fire cues were tested on the seeds of this species, which were collected from a mountainous subtropical area of Mexico (see C. *longipedicellata*; Zuloaga-Aguilar et al. 2011). Heat shock treatment, when applied on its own, significantly improved germination, while smoke water and ash had no significant effect. Combinations of heat shock with smoke water and/or ash treatments also significantly promoted germination.

Crotalaria pallida **Ait.**

SMOOTH RATTLEBOX

Exposure to aerosol smoke for 10 and 30 minutes inhibited seed germination in this upland Florida plant species (Lindon and Menges 2008).

Cullen tenax **(Lindl.) J. W. Grimes**

TOUGH SCURF-PEA

This species occurs in the eastern states of Australia. Smoke water treatments of seeds collected from the grassy woodlands of the New England Tableland of New South Wales had no effect on its germination (Clarke et al. 2000).

Cyclopia intermedia **E. Mey.**

HONEY BUSH TEA

Honey bush tea is a native of the Cape Floristic Region of South Africa. Treatment with aerosol smoke significantly improved germination (Brown et al 1995; Brown et al. 2003; Brown and Botha 2004).

Dalbergia latifolia **Roxb.**

BOMBAY BLACKWOOD

In this study, the seeds of *D. latifolia* were collected from a dry deciduous forest in India and later exposed to aerosol smoke until they turned completely brown. The seeds were arranged onto filter papers that were hung in fumigation tents, in which litter collected from locally grown dry deciduous trees was burned to generate the smoke. Germination percentages and the germination velocity index of seeds were measured (Singh and Raizada 2010). The results revealed that both germination percentage (62%) and germination velocity index (3.53) significantly increased following the smoke treatment. This was compared to 37% and 1.94, respectively, for the control group.

Dalea purpurea **Vent.**

PURPLE PRAIRIE CLOVER

Purple prairie clover seed, collected from a tallgrass prairie in the Midwest of the United States, did not germinate in response to aerosol smoke treatments of 1, 10, or 60 minutes duration (Jefferson et al. 2007).

Daviesia buxifolia **Benth.**

BOX LEAF BITTER PEA

Soil samples from the Eden Burning Study Area, a dry sclerophyll forest in the Yalumba State Forest of New South Wales, Australia, were collected and air dried to test the effects of heat, smoke, and an interaction between the two cues on seeds from the seed bank. Samples exposed to heat treatment were incubated at 80°C for 60 minutes while those exposed to smoke were incubated in a room, where smoke was generated for 120 minutes. Only the heat treatment significantly improved the germination of the seeds of this species (Penman et al. 2008). Neither smoke nor the interaction between the two cues had any effect.

Daviesia genistifolia **A. Cunn. ex Benth.**

BROOM BITTER-PEA

Broom bitter pea occurs in the eastern states of Australia. Smoke water treatments of seeds collected from the grassy woodlands of the New England Tableland of New South Wales, Australia, had no effect on its germination (Clarke et al. 2000).

Daviesia ulcifolia **Andrews**

GORSE BITTER PEA

This species occurs throughout most of Australia. Aerosol smoke treatments of 60 minutes duration applied to soil seed banks in an open-cut coal mine of the Hunter Valley of New South Wales, Australia, did not significantly improve or promote seed germination (Read et al. 2000).

Desmodium varians **(Labill.) G. Don**

SLENDER TICK-TREFOIL

Desmodium varians occurs in the eastern states of Australia. Smoke water treatments of seeds collected from the grassy woodlands of the New England Tableland of New South Wales, Australia, had no effect on its germination (Clarke et al. 2000).

Dillwynia **sp.**

The effects of both heat and smoke on the germination of soil-stored seeds of an unidentified species of *Dillwynia*, which occurs in southeastern Australian sand heathland, were researched by Wills and Read (2002). Smoke had no effect on seed germination.

Dorycnium pentaphyllum **Scop.**

PROSTRATE CANARY CLOVER

The seeds of this species, collected from the Mediterranean Basin, were incubated for 24 hours in two smoke water solutions of 1:1 and 1:10 concentrations, prepared according to Jager et al. (1996b). The smoke water solutions had no effect on germination percentage or rate, but did affect seedling growth (Moreira et al. 2010).

Genista scorpius **Georgi**

BROOM

Like *D. pentaphyllum* above, the seeds of this species were also collected from the Mediterranean Basin and incubated for 24 hours in two smoke water solutions to test the effects of smoke on germination (Moreira et al. 2010). Neither of the smoke water solutions had any effect on germination percentage or rate and did not affect seedling growth.

Genista umbellata **(L'Her) Poir**

BOLINA

Like *G. scorpius* above, the seeds of this species were also collected from the Mediterranean Basin and incubated for 24 hours in two smoke water solutions to test the effects of smoke on germination (Moreira et al. 2010). Neither of the smoke water solutions had any effect on germination percentage or rate and did not affect seedling growth.

Glycine **sp.**

GLYCINE

Aerosol smoke treatments of 60 minutes duration applied to soil seed banks in an open-cut coal mine of the Hunter Valley in New South Wales, Australia, and containing an unidentified species of *Glycine*, did not significantly improve or promote its germination (Read et al. 2000).

Glycine tomentella **Hayata**

WOOLLY GLYCINE

Woolly glycine occurs in the eastern states Australia and in parts of Asia. Smoke water treatments of seeds collected from the grassy woodlands of the New England Tableland of New South Wales, Australia, had no effect on its germination (Clarke et al. 2000).

Gompholobium knightianum **Lindl.**

Gompholobium knightianum is native to the proteaceous heathland, jarrah forest and banksia woodland regions of southwestern Western Australia. Interactions between high temperatures, aerosol smoke, and time did not significantly improve germination for this species even though Roche et al. (1997a) reported that smoke did significantly promote it in glasshouse trials and in the field.

Gompholobium marginatum **R. Br.**

RAINBOW PEA

Rainbow pea is native to the proteaceous heathland, jarrah forest, and banksia woodland regions of the southwestern region of Western Australia. The germination of this legume responded positively to an in situ aerosol smoke treatment of the soil surface (Roche et al. 1997a). However, Tieu et al. (2001a) reported that smoke was not as effective as heat treatments. Roche et al. (1997a) reported that smoke water treatments had no effect on germination.

Gompholobium preissii **Meisn.**

Like *G. knightianum*, a 1 L/m² application of undiluted smoke water to the soil surface of a rehabilitated bauxite mine in the southwest of Western Australia had no significant effect on the germination of seeds from the soil seed bank (Roche et al. 1997a).

Gompholobium tomentosum **Labill.**

HAIRY YELLOW PEA

Gompholobium tomentosum occurs throughout the southwestern region of Western Australia. Smoke water sprays, with concentrations of 50 and 100 mL/m², did not significantly promote germination of seeds occurring naturally in an intact banksia woodland 20 km south of Perth, Western Australia (Lloyd et al. 2000).

Hardenbergia violacea **(Schneev.) Stearn**

NATIVE SARSAPARILLA

This species is common in the eastern and southern parts of Australia. Smoke water treatments of seeds collected from the grassy woodlands of the New England Tableland of New South Wales had no effect on its germination (Clarke et al. 2000).

Hovea chorizemifolia **(Sweet) DC.**

HOLLY-LEAVED HOVEA

Germination of *H. chorizemifolia*, a native to Western Australia's proteaceous heathland, jarrah forest, and banksia woodland communities, was significantly promoted by smoke water and aerosol smoke applications to the soil seed bank on a bauxite mine site undergoing rehabilitation and within an undisturbed soil profile in the surrounding jarrah forest (Roche et al. 1997a). Ex situ germination trials yielded similar results (Roche et al. 1997b).

Hovea linearis **(Sm.) R. Br.**

COMMON HOVEA

Hovea linearis occurs in the Australian states of Queensland and New South Wales. Smoke water treatment of seeds collected from the grassy woodlands of the New England Tableland of New South Wales had no effect on its germination (Clarke et al. 2000).

Hovea pungens **Benth.**

DEVIL'S PINS

This species occurs in the southwestern region of Western Australia. Freshly collected seeds were treated for 60 minutes with cool aerosol smoke, but did not significantly respond to the treatment (Roche et al. 1997b).

Hovea trisperma **Benth.**

COMMON HOVEA

Common hovea occurs throughout the Southwest Botanical Province of Western Australia. Germination of its seeds was significantly increased when treated with aerosol smoke for 60 minutes (Roche et al. 1997b).

Indigofera australis **Willd.**

AUSTRAL INDIGO

This species is common throughout Australia. Smoke water treatments of seeds collected from the grassy woodlands of the New England Tableland of New South Wales had no effect on its germination (Clarke et al. 2000).

Indigofera filifolia **Thunb.**

Seed germination in *Indigofera filifolia*, a native of the Cape Floristic Region of South Africa, was not significantly promoted after 30 minutes of aerosol smoke treatment (Brown et al. 2003), but did germinate readily in response to hot water (Brown and Botha 2004).

Jacksonia scoparia **R. Br.**

DOGWOOD

This species occurs in the Australian states of Queensland and New South Wales. Smoke water treatments of seeds collected from the grassy woodlands of the New England Tableland of New South Wales had no effect on its germination (Clarke et al. 2000).

Kennedia prostrata **R. Br.**

RUNNING POSTMAN

The running postman is common throughout southern Australia. A 1 L/m² application of undiluted smoke water to the soil surface of a rehabilitated bauxite mine in
the southwest of Western Australia had no significant effect on the germination of
seeds from the soil seed bank (Roche et al. 1997a). See also *K. coccinea* on page 209.

Lespedeza bicolor **Turcz.**

SHRUBBY BUSHCLOVER

The effects of aerosol smoke, heat, darkness, cold stratification, and combinations
of smoke with each of the three other treatments on seed germination were examined in this study (Tsuyuzaki and Miyoshi 2009). Smoke was produced by burning
Timothy hay (*Phleum pratense*), which was pumped through a 3.5 m cooling tube
into a smoke chamber for approximately 5 minutes. The seeds were exposed to the
smoke for 60 minutes. Those seeds exposed also to heat were incubated at 75°C for
25 minutes. The cold stratification process took 1 month, during which the seeds
remained in an incubator set to 4°C. Where the dark treatment was concerned, the
seeds were maintained in total darkness for the entire germination period. Germination in both seed batches responded positively to the heat treatment, but did not
respond to any of the other treatments. Specifically, when the 2004 seed batch was
treated with heat, there was 16% germination versus 2% for the control group. In
all, 38% of the seeds of the 2005 batch germinated following heat treatment. Only
4% of the control group germinated for that batch.

Lespedeza capitata **Michx.**

ROUND HEAD LESPEDEZA

Germination of *L. capitata*, native to the eastern and central United States and
central Canada, increased to 80% when its seeds were treated with 1, 10, or 60
minutes of aerosol smoke (Jefferson et al. 2007). Without treatment, only 50% of
the seeds germinated.

Lespedeza cuneata **(Dum. Cours) G. Don**

CHINESE BUSHCLOVER

Exposure to aqueous smoke extracts of various concentrations for 10 or 45 minutes duration did not significantly increase seed germination of Chinese bushclover seeds (Smith 2006).

Lespedeza juncea **(L. f.) Pers. ssp.** Juncea

BUSH CLOVER

Smoke water treatments of seeds collected from the grassy woodlands of the New
England Tableland of New South Wales had no effect on its germination (Clarke
et al. 2000).

Lotus australis **Andrews**

AUSTRAL TREFOIL

This species is common throughout Australia and parts of Asia. Smoke water treatments of seeds collected from the grassy woodlands of the New England Tableland of New South Wales had no effect on its germination (Clarke et al. 2000).

Lotus corniculatus **L.**

BIRDSFOOT TREFOIL

Smoke water solutions at various concentrations and exposures of 10 and 45 minutes did not significantly improve germination of the seeds of *L. corniculatus* (Smith 2006). High concentrations of smoke water solutions, when applied for 10 minutes, inhibited germination. This species is native to parts of Eurasia and northern Africa.

Lotus corniculatus **L. var.** Japonica

JAPANESE BIRDSFOOT TREFOIL

The effects of aerosol smoke, heat, darkness, cold stratification, and combinations of smoke with each of the three other treatments on seed germination were examined in this study (Tsuyuzaki and Miyoshi 2009; for details about the tests performed, see *Lespedeza bicolor* above). None of the treatments had any effect on the germination of this species.

Lotus salsuginosus **Greene**

COASTAL BIRD'S-FOOT TREFOIL

This lotus is commonly found growing in the southwest of the United States and northern Mexico. Germination of *L. salsuginosus* was inhibited by treatment with charate (0.5 g per Petri dish) (Keeley and Keeley 1987).

Lotus scoparius **(Nutt.) Ottley**

COMMON DEER WEED

Lotus scoparius commonly occurs in the chaparrals of the southwest of the United States and northern Mexico. Germination increased when its seeds were treated with heat (120°C for 5 minutes). The addition of charate (0.25 g on filter paper) after the heat treatment had an inhibitory effect on germination. Charate alone had no effect (Keeley 1987).

Lotus strigosus **(Nutt.) Greene**

STRIGOSE BIRD'S-FOOT TREFOIL

This lotus also commonly occurs in the southwest of the United States and northern Mexico. Germination of its seed was inhibited by treatment with charate (0.5 g per Petri dish) (Keeley and Keeley 1987).

Lupinus exaltatus **Zucc.**

LUPINUS

A variety of different fire cues were tested on the seeds of this species, which were collected from a mixed forest located in a mountainous subtropical area of Mexico (Zuloaga-Aguilar et al. 2011). These included heat shock (100:15 sand: water [w:w] substrate at 120°C for 5 minutes), soaking the seeds in smoke water for 3 hours (prepared by burning 150 g of mixed forest litter and bubbling the resultant smoke into 1.5 L of distilled water, and adjusting the pH to 5 using sodium hydroxide) or ash (1.5 g of fine ash was added to the agar plates used to germinate the seeds). Combinations of these treatments were also tested for their effects on seed germination. Heat shock significantly increased germination percentage of this species, while the smoke water and ash treatments had no significant effect. Combinations of heat shock with the smoke water and heat shock/ash treatments also significantly improved germination.

Medicago sativa **L.**

ALFALFA

A combination of aerosol smoke treatments of 15, 30, or 45 minutes duration, in combination with three different treatments of dissolved aspirin (0.145wt%; 10, 20, and 30 minutes; Bayer) increased both the germination of alfalfa seeds and their growth ratio (Hong and Kang, 2011).

Mimosa galeottii **Benth.**

TALL THIMBLEWEED

Like *Lupinus exaltatus* above, a variety of different fire cues were tested on the seeds of this species, which were collected from a mixed forest located in a mountainous subtropical area of Mexico (see *L. exaltatus* for details; Zuloaga-Aguilar et al. 2011). The heat shock treatment, when applied on its own, significantly increased germination percentages for this species. The smoke water and ash treatments had no significant effect. The heat shock/smoke water treatment and the heat/ash combination both significantly improved germination.

Mirbelia dilatata **R. Br.**

HOLLY-LEAVED MIRBELIA.

This species, like the genus it belongs to, is endemic to Australia. A 1 L/m² application of undiluted smoke water to the soil surface of a rehabilitated bauxite mine in the southwest of Western Australia had no significant effect on the germination of seeds from the soil seed bank (Roche et al. 1997a).

Ononis minutissima **L.**

ANONIS MENOR

The seeds of this species were incubated for 24 hours in two smoke water solutions of 1:1 and 1:10 concentrations, prepared according to Jager et al. (1996b). Neither of the smoke water solutions had any effect on germination or seedling growth (Moreira et al. 2010). This species occurs in the Mediterranean Basin.

Otholobium fruticans **(L.) C. H. Stirt.**

CAPETOWN PEA

Germination of this vulnerable (IUCN 2013) South African fynbos legume, which occurs in the Cape Floristic Region, significantly increased in response to treating the seed with aerosol smoke. Germination without the treatment was 4%, compared with 65% using aerosol smoke treatment (Brown and Botha 2004). Brown et al. (2003) reported there were mixed results when seeds were treated with aerosol smoke.

Podalyria calyptrata **Willd.**

SWEET PEA BUSH

The sweet pea bush occurs in the Cape Floristic Region of South Africa and elsewhere in the country. Aerosol smoke had no effect on its seed germination (Brown et al. 2003).

Podalyria sericea **(Andrews) R. Br.**

LESSER BUSH SWEET PEA

Seed germination in *P. sericea*, a native of the Cape Floristic Region of South Africa, was not significantly promoted after 30 minutes of aerosol smoke treatment (Brown et al. 2003), but did germinate readily in response to hot water (Brown and Botha 2004).

Prosopis glandulosa **Torr.**

HONEY MESQUITE

Smoke water, heat, and combinations of both heat and smoke were tested on the seeds of this species by Chou et al. (2012). The seeds were soaked for 20 hours in the commercially available Regen 2000® smoke water solution, at concentrations of 1:5, 1:10, or 1:100 (v/v), after which they were exposed to heat treatments of 50 or 80°C for 5 minutes. Neither germination percentage nor mean germination times were significantly affected by those treatments. The seeds did not respond to heat treatments and there was no significant interaction between heat and smoke.

Psoralea pinnata **L.**

FOUNTAIN BUSH

Fountain bush is a native of the Cape Floristic Region of South Africa. Thirty minutes of aerosol smoke treatment did not significantly improve seed germination (Brown et al. 2003). Hot water did, however, promote germination (Brown and Botha 2004).

Pultenaea microphylla **Sieber ex DC.**

Pultenaea microphylla grows in the dry sclerophyll woodlands of eastern Australia. Clarke et al. (2000) showed that germinating seeds in Petri dishes moistened with 10% smoke water significantly inhibited germination.

Retama sphaerocarpa **(L.) Boiss.**

COMMON RETAMA

Common retama is native to the grasslands and open woodlands of southeastern Europe and northern Africa. Germination of this species was increased to approximately 20% by treating the seeds with a 10-minute exposure to aerosol smoke (Pérez-Fernández and Rodríguez-Echeverría 2003).

Sphaerolobium medium **R. Br.**

The species, like its genus, is endemic to Australia. A 1 L/m² solution of undiluted smoke water applied to the soil surface of a rehabilitated bauxite mine in the southwest of Western Australia had no significant effect on the germination of seeds from the soil seed bank (Roche et al. 1997a).

Tephrosia pedicellata **Baker**

SCARLET HAWTHORN

Seeds collected from a Sudanian savanna-woodland in Burkina Faso, Africa, were treated with a variety of fire cues to determine their effects on seed germination (Dayamba et al. 2010). The seeds were soaked in smoke water (at concentrations of 100%, 75%, 50%, 25%, and 5% of the stock solution) for 24 hours. The smoke water stock solution was produced by burning a mixture of dominant native species from the Tiogo and Laba State forests of Sudan, and pumping the smoke through water for 10 hours. The seeds also underwent a heat shock treatment, during which they were incubated in an oven at 40, 80, 120, or 140°C for 2.5 minutes. Following these treatments, germination percentages and mean germination times were measured. None of the smoke or heat shock treatments had any effect on the germination of this species.

Trifolium ambiguum **M. Bieb.**

KURA CLOVER

Aqueous smoke treatments (at 10 and 45 minute exposure and at various concentrations) of the seeds of kura clover ("perfect fit" variety) did not significantly improve germination (Smith 2006). High concentrations of smoke water solutions, when applied for 10 minutes, increased time to germination (T_{50}) to 50%. *Trifolium ambiguum* is native to both Asia and Australia.

Trifolium angustifolium **L.**

NARROWLEAF CRIMSON CLOVER

Native to southern Europe, western Asia and northern Africa, germination of this clover significantly increased when seeds were exposed to 10 minutes of aerosol smoke. Maximum germination of the treated seeds was 30%–40% (Pérez-Fernández and Rodríguez-Echeverría 2003).

Virgilia divaricata **Adamson**

BLOSSOM TREE

Blossom tree is a native of the Cape Floristic Region of South Africa. Thirty minutes of aerosol smoke treatment did not significantly improve seed germination (Brown et al. 2003), but was significantly promoted when treated with hot water (Brown and Botha 2004).

Ulex borgiae **Rivas Mart.**

GORSE

The seeds of this species, collected from the Mediterranean Basin, were incubated for 24 hours in two smoke water solutions of 1:1 and 1:10 concentrations (as according to Jager et al. 1996b). Neither of the smoke water solutions had any effect on germination or seedling growth (Moreira et al. 2010).

Ulex parviflorus **Pourr.**

SMALL-FLOWERED GORSE

Like *U. borgiae* above, the seeds of this species were also collected from the Mediterranean Basin and incubated for 24 hours in two smoke water solutions of 1:1 and 1:10 concentrations (as according to Jager et al. 1996b). Neither of the smoke water solutions had any effect on germination or seedling growth (Moreira et al. 2010).

FAGACEAE

Quercus ilex **L. ssp.** ballota **(Desf.) Samp.**

HOLM OAK

The effects of aerosol smoke (5, 10, and 15 minutes exposure; prepared according to Baxter et al. 1994), heat at 60°C for 5 and 15 minutes, as well as 90, 110, or 150°C for 5 minutes, ash (0.168 g of ash to each incubation tray), ash solution (120 mL of ash in distilled water with 5 g/L applied to the seeds), and charcoal (1 g of fragmented charcoal per tray) were all tested on the seeds of this species (Reyes and Casal 2006). Germination percentages and rates were recorded after a period of 22 weeks and then again at 1 year. After 22 weeks, germination rates were significantly decreased, as was the final germination percentage, following 5 minutes of heat shock treatment at 150°C. None of the other treatments had any effect on germination.

Quercus pyrenaica **L.**

PYRENEAN OAK

The seeds of this oak were exposed to a number of different fire cues, to assess their effects on germination percentage and rate (Reyes and Casal 2006; see

Quercus ilex ssp. *ballota* above for details about the tests performed). After 22 weeks, the smoke treatment, the ash dilution, and the ash and charcoal treatments significantly increased germination rates. After 1 year, germination rates appeared to have increased for seeds treated with aerosol smoke for 5 minutes, as well as the ash and ash dilution treatments, but were not significant. A smoke application trial of 10 minutes resulted in decreased germination rates, but, again, was not significantly different. None of the treatments had any effect on germination percentages.

Quercus robur **L.**

ENGLISH OAK

The seeds of this oak were exposed to a number of different fire cues, to assess their effects on germination percentage and rate (Reyes and Casal 2006; see *Quercus ilex* ssp. *ballota* above for details about the tests performed). The germination rates at the 22-week mark decreased significantly after they were exposed to 5 minutes of aerosol smoke. After 1 year, the germination rates of seeds following the treatment were not significantly different compared to those of the control. At 22 weeks, smoke treatment of 5 minutes duration resulted in a significantly decreased germination percentage. The aerosol smoke treatment of 10 minutes promoted a significantly greater germination percentage. None of the treatments had any effect on the germination percentage after 1 year.

GARRYACEAE

Garrya flavescens **S. Watson**

ASHY SILKTASSEL

Ashy silktassel is widely distributed across the southwest of the United States and northern Mexico. Charate significantly promoted germination of this species. Four percent germination was observed in the control, versus 65% germination when seeds were treated with charate (Keeley 1987).

GENTIANACEAE

Centaurium erythraea **Rafn**

COMMON CENTAURY

Soil samples from the Eden Burning Study Area, a dry sclerophyll forest in the Yalumba State Forest of New South Wales, Australia, were collected and air dried to test the effects of heat, smoke, and an interaction between the two cues on seeds from the seed bank. Samples exposed to heat treatment were incubated at 80°C for 60 minutes while those exposed to smoke were incubated in a room, where smoke was generated for 120 minutes. None of the treatments had any effect on the germination of the seeds of this species (Penman et al. 2008).

Chironia linoides **L. ssp.** emarginata **(Jaroscz) I.Verd**

This species occurs in the Cape Floristic Region of South Africa and elsewhere in Africa, including Lesotho. Aerosol smoke had no effect on seed germination (Brown et al. 2003).

GERANIACEAE

Geranium incanum **Burm. f.**

CARPET GERANIUM

Brown and Botha (2004) reported that carpet geranium, a native of South Africa, responded positively to aerosol smoke treatment. Germination increased from 60% in the control to 85% when seeds had been exposed to smoke.

Pelargonium auritum **Willd.**

This species occurs in the Cape Floristic Region of South Africa. Aerosol smoke had no effect on its seed germination (Brown et al. 2003).

Pelargonium capitatum **(L.) L'Hér. ex Ait.**

ROSE-SCENTED GERANIUM

Rose scented geranium is native to South Africa. The species germinated more readily when seeds were exposed to aerosol smoke for 60 minutes (Brown et al. 1995). Brown et al. (2003) later reported that aerosol smoke had no significant effect on seed germination.

Pelargonium crithmifolium **Sm.**

SAMPHIRE LEAFED GERANIUM

Germination of this South African geranium doubled in response to aerosol smoke treatment (Brown et al. 1995; Brown et al. 2003; Brown and Botha 2004).

Pelargonium cucullatum **(L.) L'Hér.**

TREE PELARGONIUM

Germination of tree pelargonium seeds increased significantly in response to smoke treatment. Germination, however, was still only 15%. Treating the seed with hot water increased germination to 59% (Brown and Botha 2004). Brown et al. (2003) earlier reported that aerosol smoke had no significant effect on seed germination. This species is indigenous to South Africa.

Pelargonium peltatum **(L.) L'Hér.**

IVY-LEAFED GERANIUM

This geranium occurs in the Cape Floristic Region of South Africa. Aerosol smoke had no effect on seed germination (Brown et al. 2003).

Pelargonium quercifolium **L'Hér.**

This geranium also occurs in the Cape Floristic Region of South Africa and else-where in southern Africa. Aerosol smoke had no effect on seed germination (Brown et al. 2003).

Pelargonium **sp.**

PELARGONIUMS

Seed germination in an unidentified species of *Pelargonium*, which occurs in the Cape Floristic Region of South Africa, was not promoted by aerosol smoke (Brown et al. 2003).

Pelargonium suburbanum **Clifford ex D. A. Boucher**

WILDEMALVA

Pelargonium suburbanum occurs in the Cape Floristic Region of South Africa and elsewhere in southern Africa. Treatment with aerosol smoke had no effect on its germination (Brown et al. 2003).

GOODENIACEAE

Brunonia australis **Sm.**

NATIVE CORNFLOWER

Brunonia australis is widely distributed throughout Australia. Germination was significantly promoted when sown seeds were treated with aerosol smoke for 60 minutes (Roche et al. 1997b). Karrikinolide treatments (10 ppb) also sig-nificantly increased the mean germination percentage of this species (Flematti et al. 2004).

Dampiera linearis **R. Br.**

COMMON DAMPIERA

Dampiera linearis is endemic to the southwestern region of Western Australia. Smoke water sprays, with concentrations of 50 and 100 mL/m^2, did not signifi-cantly promote germination of common dampiera occurring naturally in an intact banksia woodland 20 km south of Perth, Western Australia (Lloyd et al. 2000). Germination of its seed was also unaffected when its seeds were soaked in 10% smoke water for 24 hours (Clarke and French 2005).

Goodenia caerulea **R. Br.**

Dixon et al. (1995) reported that 90 minutes of cold smoke treatment had no effect on seed germination of this Western Australian species. A 60 minute exposure to cold smoke was also ineffective (Roche et al. 1997b).

Goodenia geniculata **de Vriese**

BENT GOODENIA

Germination of bent goodenia seeds, collected from a *Eucalyptus baxteri* heathy-woodland in Victoria, Australia, significantly improved following a combined heat and smoke water treatment (Enright and Kintrup 2001).

Lechenaultia biloba **Lindl.**

BLUE LECHENAULTIA

Lechenaultia biloba is a native of the proteaceous heathland, banksia woodland, and jarrah forest communities of the southwestern region of Western Australia. Aerosol smoke applications significantly increased the in situ and ex situ germination of this species (Dixon et al. 1995; Roche et al. 1997a), but smoke water treatments had no effect (Roche et al. 1997a). Norman et al. (2006) reported that neither aerosol smoke nor smoke water treatments significantly promoted seed germination.

Lechenaultia floribunda **Benth.**

FREE-FLOWERING LECHENAULTIA

Lechenaultia floribunda is a native of southwestern Western Australia's proteaceous heathland, banksia woodland, and jarrah forest regions. Ex situ germination was significantly improved when the soil seed bank was treated with aerosol smoke for 90 minutes (Dixon et al. 1995).

Lechenaultia formosa **R. Br.**

RED LECHENAULTIA

Red lechenaultia occurs commonly in the proteaceous heathland, mixed eucalypt woodland and jarrah forest regions of the southwest of Western Australia. Germination of this species was significantly increased by an aerosol smoke treatment of the soil surface in which the seed had been sown (Dixon et al. 1995).

Lechenaultia macrantha **K. Krause**

WREATH LECHENAULTIA

Wreath lechenaultia is a native of southwestern Western Australia's proteaceous heathland, banksia woodland, and jarrah forest regions. Ex situ germination was significantly improved when the soil seed bank was treated with aerosol smoke for 90 minutes (Dixon et al. 1995).

Scaevola calliptera **Benth.**

Germination of *S. calliptera*, which is native to Western Australia's proteaceous heathland, banksia woodland, and jarrah forest communities, increased in response to ex situ aerosol smoke (Dixon et al. 1995; Roche et al. 1997b) and in situ smoke water application of the soil seed bank on a rehabilitation area of a bauxite mine site in Western Australia (Roche et al. 1997a).

Scaevola crassifolia **Labill.**

THICK-LEAVED FAN-FLOWER

This thick-leaved fan-flower grows in the coastal regions of western and southern Australia. Germination was significantly increased once sown seeds were exposed to aerosol smoke for 60 minutes (Roche et al. 1997b).

Scaevola fasciculata **Benth.**

BRISTLY SCAEVOLA

Bristly scaevola occurs in the Darling Range and adjacent areas of the southwestern region of Western Australia. The best germination results for this species were achieved when the seeds were stored in soil for 12 months and then treated with aerosol smoke for 60 minutes (Roche et al. 1997b; Tieu et al. 2001b).

Scaevola paludosa **R. Br.**

Scaevola paludosa occurs in limited areas of the southwest of Western Australia. Smoke water sprays, with concentrations of 50 and 100 mL/m^2, did not significantly promote germination of seeds collected from an intact banksia woodland 20 km south of Perth, Western Australia (Lloyd et al. 2000).

Scaevola pilosa **Benth.**

This species occurs in the southwestern region of Western Australia. Freshly collected seeds of this Australian plant species, when treated with 60 minutes of cool aerosol smoke, did not significantly respond to the treatment (Roche et al. 1997b).

Scaevola thesioides **Benth.**

Scaevola thesioides commonly occurs in the coastal areas and adjacent sandplains of the Southwest Botanical Province of Western Australia. Germination was significantly promoted when seeds were initially immersed in karrikinolide (10 ppb) for 24 hours prior to sowing the seeds in soil (Flematti et al. 2004).

Velleia paradoxa **R. Br.**

SPUR VELLEIA

This species is common throughout the eastern states of Australia. Smoke water treatments of seeds collected from the grassy woodlands of the New England Tableland of New South Wales, Australia, had no effect on its germination (Clarke et al. 2000).

Velleia rosea **S. Moore**

Dixon et al. (1995) reported that 90 minutes of cold smoke treatment had no effect on seed germination of this Western Australian species.

Velleia trinervis **Labill.**

Velleia trinervis is widely distributed throughout the Southwest Botanical Province of Western Australia. Germination was significantly promoted when sown seeds were exposed to aerosol smoke for 60 minutes (Roche et al. 1997b). Smoke water (1%) and aerosol smoke (60 minutes) was also shown to promote germination when the seeds were sown on a bauxite mine rehabilitation area (Norman et al. 2006).

GYROSTEMONACEAE

Codonocarpus cotinifolius **(Desf.) F. Muell.**

Codonocarpus cotinifolius is a native of Western Australia's proteaceous heathland, banksia woodland, and jarrah forest regions. Ex situ germination improved when the soil seed bank was exposed to aerosol smoke for 90 minutes (Dixon et al. 1995). Smoke water treatment, with or without heat (70°C for 60 minutes), also significantly improved germination in this species (Baker et al. 2005).

Gyrostemon racemiger **H. Walter**

Smoke water treatment of seeds collected from Gin Gin, Western Australia, had no effect on the germination of this species, even when used in combination with other treatments, such as heat and manual scarification (Baker et al. 2005). In another study, the effects of "Seed Starter" smoke water, karrikinolide (KAR$_1$), glyceronitrile and cellulose-derived smoke water were tested on the seeds of this species (Downes et al. 2013). Germination was promoted by "Seed Starter" smoke water and to some degree by the cellulose-derived smoke water, but was not affected by KAR$_1$ or glyceronitrile, suggesting that other chemicals in smoke promote germination. *Gyrostemon racemiger* is native to Western Australia.

Gyrostemon ramulosus **Desf.**

CORKY BARK

Gyrostemon ramulosus is a native of Western Australia's proteaceous heathland, banksia woodland, and jarrah forest regions. Ex situ germination significantly improved when the soil seed bank was exposed to aerosol smoke for 90 minutes (Dixon et al. 1995). Like *G. racemiger*, the effects of "Seed Starter" smoke water, karrikinolide (KAR$_1$), glyceronitrile and cellulose-derived smoke water were also tested on the seeds of the corky bell fruit (Downes et al. 2013). While germination was promoted by "Seed Starter" smoke water and to a lesser degree by the cellulose-derived smoke water, KAR$_1$ and glyceronitrile had no effect. *Gyrostemon ramulosus* is native to Western Australia.

Tersonia cyathiflora **(Fenzl) A. S. George**

BUTTON CREEPER

Button creeper is native to the proteaceous heathland and banksia woodland of the west coast of Western Australia. The best results for seed germination occurred when the seeds were stored in soil for 12 months, sown and then treated with an aerosol smoke application of 60 minutes duration (Roche et al. 1997b). Baker et al. (2005) reported that germination increased when seeds were stored under natural conditions (burial of seeds in the soil) for 24 months and then incubated in Petri dishes containing smoke water (10%). An average of 80%–90% of seeds germinated under those conditions, compared to less than 5% when they received no treatment. In another study, seeds collected from Lesueur National Park, Western Australia, were treated with smoke water (1:10 [v/v] dilution of smoke water; pH = 4.05, "Seed Starter," Kings Park and Botanic Garden, Perth, Western Australia) and to karrikinolide (KAR$_1$; 1 µM) by Downes et al. (2010), who reported that the smoke water significantly increased germination but the seeds did not respond to karrikinolide. The authors concluded that other chemicals in the smoke may have promoted germination.

HALORAGACEAE

Glischrocaryon aureum **(Lindl.) Orchard**

COMMON POPFLOWER

This species commonly occurs in the jarrah (*Eucalyptus marginata*) forests of Western Australia. Neither aerosol smoke nor smoke water treatments significantly promoted germination of its seeds (Roche et al. 1997a; Norman et al. 2006).

Gonocarpus cordiger **(Fenzl) Nees**

This species also occurs in the jarrah forests of Western Australia. Neither aerosol smoke nor smoke water treatments significantly promoted germination (Roche et al. 1997b; Norman et al. 2006).

Gonocarpus micranthus **Thunb.**

CREEPING RASPWORT

Soil samples from the Eden Burning Study Area, a dry sclerophyll forest in the Yalumba State Forest of New South Wales, Australia, were collected and air dried to test the effects of heat, smoke, and an interaction between the two cues on seeds from the seed bank. Samples exposed to heat treatment were incubated at 80°C for 60 minutes while those exposed to smoke were incubated in a room, where smoke was generated for 120 minutes. Only the heat treatment significantly improved the germination of the seeds of this species (Penman et al. 2008). Neither smoke nor the interaction between the two cues had any effect.

Gonocarpus pithyoides **Nees**

This species occurs in banksia woodlands in Western Australia. Germination of its seeds was not significantly promoted when they were soaked in 10% smoke water for 24 hours (Clarke and French 2005).

Gonocarpus teucrioides **DC.**

GERMANDER RASPWORT

Soil samples from the Eden Burning Study Area, a dry sclerophyll forest in the Yalumba State Forest of New South Wales, Australia, were collected and air dried to test the effects of heat, smoke, and an interaction between the two cues on seeds from the seed bank. Samples exposed to heat treatment were incubated at 80°C for 60 minutes while those exposed to smoke were incubated in a room, where smoke was generated for 120 minutes. Only the smoke treatment significantly improved the germination of the seeds of this species (Penman et al. 2008). Neither heat nor the interaction between the two cues had any effect.

Emmenanthe penduliflora

HYDROPHYLLACEAE

Emmenanthe penduliflora **Benth.**

WHISPERING BELLS

Whispering bells are native to the chaparral and desert regions of the southwestern of the United States. This species does not germinate without its seeds first being treated with charred wood leachate (Wicklow 1977; Jones and Schlesinger 1980; Keeley et al. 1985), 5 minutes of aerosol smoke (Fotheringham et al. 1995; Keeley and Fotheringham 1998a), smoke water dilutions (Keeley and Fotheringham 1998a), or karrikinolide (10 ppb) (Flematti et al. 2004). Flematti et al. (2007) reported that a number of analogues of karrikinolide were effective in also promoting germination in this species.

Eriodictyon crassifolium **Benth.**

THICKLEAF YERBA SANTA

Eriodictyon crassifolium is endemic to the Californian chaparrals of the United States. Keeley (1987) discovered that charate significantly promoted germination of this species. Germination increased significantly when seeds were first given a heat shock treatment (100°C or 120°C for 5 minutes) prior to the charate application. A similar response was achieved with combined smoke and heat treatments (Keeley et al. 2005).

Eucrypta chrysanthemifolia **(Benth.) Greene**

SPOTTED HIDESEED

Spotted hideseed is native to the Californian chaparral of the United States. Keeley and Fotheringham (1998a) reported that germination of this species increased in response to aerosol smoke treatment.

Phacelia cicutaria **E. Greene**

CATERPILLAR PHACELIA

Germination of this native of the Californian chaparral and the Great Basin in Nevada, United States, significantly increased when powdered charred wood was applied to its seeds (Keeley 1984; Keeley et al. 1985).

Phacelia fremontii **Torr.**

FREMONT'S PHACELIA

This species is native to the southwestern region of the United States. Globally and locally, this species is considered rare (IUCN 2013). Germination of its seeds was slightly improved using a seed pretreatment of powdered charred wood. This increased germination from 4% to 11% (Keeley et al. 1985).

Phacelia grandiflora **(Benth.) A. Gray**

LARGE FLOWER PHACELIA

Largeflower phacelia is native to the Californian chaparrals of the United States. Germination was improved when powdered charred wood was applied to its seeds (Keeley et al. 1985). Keeley and Fotheringham (1998a) showed that 5 or 8 minutes exposure to aerosol smoke also promoted germination.

Phacelia minor **(Harv.) Thell.**

WILD CANTERBURY BELLS

Wild canterbury bells is native to the Californian chaparrals of the United States. Keeley and Keeley (1987) reported that charate treatment of its seeds improved the germination response of this species from 0% to 13%. Keeley and Fotheringham (1998a) also demonstrated that germination of this species was significantly increased in response to aerosol smoke treatment.

Phacelia strictiflora **(Engelm. & A. Gray) A. Gray.**

PRAIRIE PHACELIA

Germination of the seeds of this native of the United States was enhanced when treated with dilute smoke solutions (Ross and Rice 2013; unpublished data).

ILLECEBRACEAE

Paronychia **sp.**

NAILWORT

This nailwort is native to the dry sclerophyll forests of eastern Australia. Read et al. (2000) reported that aerosol smoke treatments of 60 minutes duration promoted germination. Germination was, however, still only 6%.

LAMIACEAE

Ajuga australis **R. Br.**

AUSTRAL BUGLE

Austral bugle has a widespread distribution in southern and eastern Australia. Germination was significantly promoted by a smoke water (10%) application (Clarke et al. 2000). Fifty-nine percent of seeds germinated when treated with smoke water compared with only 19% without treatment.

Hemiandra pungens **R. Br.**

SNAKE BUSH

Aerosol smoke treatments of broadcast snake bush seeds used in rehabilitated mine sites in the southwest of Western Australia had no effect on their germination (Roche et al. 1997a).

Hemiandra **sp.**

SNAKEBUSHES

This rare and endangered snakebush is native to Western Australia's protaceous heathland, banksia woodland, and jarrah forest communities. A combination of seed pretreatments resulted in high germination, especially after partial seed coat removal, adding gibberellic acid to the growth medium, and soaking seeds in concentrated smoke water for 24 hours (Cochrane et al. 2002).

Hemigenia ramosissima **Benth.**

This rare *Hemigenia* species is native to Western Australia's jarrah forest ecosystems. In situ application of aerosol smoke to the soil surface significantly promoted its germination (Roche et al. 1997a).

Hemigenia rigida **Benth.**

This species is also native to Western Australia's jarrah forest ecosystems. Neither aerosol smoke nor smoke water treatments significantly promoted germination of its seeds (Norman et al. 2006).

Hemigenia sericea **Benth.**

SILKY HEMIGENIA

Neither aerosol smoke nor smoke water treatments significantly promoted germination in the seeds of this Western Australian species (Norman et al. 2006).

Hemiphora elderi **(F. Muell.) F. Muell.**

RED VELVET

Freshly collected seeds of this Australian plant species, when treated with 60 minutes of cool aerosol smoke, did not significantly respond to the treatment (Roche et al. 1997b).

Lachnostachys eriobotrya **(F. Muell.) Druce**

LAMBS TAILS

Dixon et al. (1995) reported that 90 minutes of cold smoke treatment had no effect on seed germination of this Western Australian species. Sixty minutes of cool aerosol smoke treatment also did not improve germination (Roche et al. 1997b).

Lachnostachys verbascifolia **F. Muell.**

LAMBS TAILS

This species occurs in Western Australia. Freshly collected seeds of this species, when treated with 60 minutes of cool aerosol smoke, did not significantly respond to the treatment (Roche et al. 1997b).

Lamium purpureum **L.**

RED DEAD NETTLE

Read dead nettle is native to Europe and considered a weed in cultivated fields in similar climatic regions. Germination was inhibited when its seeds were treated with smoke water (10% dilution) (Adkins and Peters 2001).

Lavandula latifolia **Medik.**

SPIKE LAVENDER

The seeds of this species, collected from the Mediterranean Basin, were incubated for 24 hours in two smoke water solutions of 1:1 and 1:10 concentrations, prepared according to Jager et al. (1996b). The smoke water solutions significantly increased germination percentage and germination rate but had no effect on seedling growth (Moreira et al. 2010).

Lavandula stoechas **L.**

TOPPED LAVENDER

Aqueous smoke solutions (24 hours of exposure) prepared from a variety of plant species did not significantly increase seed germination or germination rate for this woody species from the Marmaris region of southwestern Turkey (Çatav et al. 2012). In another study by Moreira et al. (2010), seeds collected from the Mediterranean Basin, were incubated for 24 hours in two smoke water solutions of 1:1 and 1:10 concentrations that were prepared according to Jager et al. (1996b). In contrast to the previous study, both of the smoke water solutions significantly increased germination percentage and germination rate but similar to the previous study had no effect on seedling growth.

Mentha satureioides **R. Br.**

CREEPING MINT

Creeping mint occurs in the eastern states of Australia. Smoke water treatments of seeds collected from the grassy woodlands of the New England Tableland of New South Wales had no effect on its germination (Clarke et al. 2000).

Microcorys eremophiloides **Kenneally**

WONGAN MICROCORYS

Wongan microcorys is a native of the proteaceous heathlands of Western Australia. The IUCN (2013) classified this species as vulnerable. A combination of seed pretreatments resulted in high germination, especially after partial seed coat removal, adding gibberellic acid to the growth medium and soaking seeds in concentrated smoke water for 24 hours (Cochrane et al. 2002).

Monarda citriodora **Cerv ex. Lag.**

LEMON BEEBALM

There was significant inhibition in germination when seeds of lemon beebalm were exposed to an aerosol smoke treatment of 8 minutes, heat treatments at 30 or 60 seconds at 100°C and wet, cold stratification for 1 month at 4°C (Schwilk and Zavala 2012). The germination of seeds stratified in a dry, cold environment at 4°C for 1 month, and with a relative humidity of 10%, was not significantly improved following smoke treatment. Chou et al. (2012) tested smoke, heat and combinations of both heat and smoke on the germination percentage and mean germination times of this species. Smoke treatments comprised of soaking seeds for 20 hours in the commercially available Regen 2000® smoke water solution, at concentrations of 1:5, 1:10, or 1:100 (v/v). The seeds were then exposed to heat treatments of 50 or 80°C for 5 minutes. None of these treatments, or interactions between them, significantly improved germination or germination times. This species is native to the United States and to Mexico.

Monarda fistulosa **L.**

WILD BERGAMOT

Wild bergamot occurs throughout Canada, the United States, and northern Mexico and has become a weed in other parts of the world. Jefferson et al. (2007)

reported that aerosol smoke inhibited germination. Sixty percent of seed germinated without treatment in comparison to 40% when seeds were exposed to aerosol smoke for 60 minutes.

Pityrodia scabra **A. S. George**

WYALKATCHEM FOXGLOVE

This species occurs in Western Australia. Freshly collected seeds of this plant, when treated with 60 minutes of cool aerosol smoke, did not significantly respond to the treatment (Roche et al. 1997b).

Plectranthus parviflorus **Willd.**

LITTLE SPURFLOWER

Little spurflower is native to eastern Australia, Hawaii, and parts of French Polynesia and Malenesia, with the species often occurring in rocky locations. Tang et al. (2003) reported that germination was promoted by aerosol smoke treatments of 60 minutes duration.

Pycnanthemum pilosum **Nutt.**

WHORLED MOUNTAIN MINT

Whorled mountain mint is native to the grasslands of eastern Canada and northeastern and northcentral United States. It is listed as rare and endangered in parts of the United States and extirpated in others. Pennacchio et al. (2005) reported that final germination of this species increased in response to 32 minutes of aerosol smoke. Seeds exposed to smoke in Petri dishes exhibited decreased final germination percentages and rate of germination, and increased mean period to final germination when treated with 1 to 4 minutes of aerosol smoke. This may be due to an inhibitor in the smoke, which is rinsed from the seeds in glasshouse studies, but not from Petri dishes.

Rosmarinus officinalis **L.**

ROSEMARY

Rosemary seeds were incubated for 24 hours in two smoke water solutions of 1:1 and 1:10 concentrations, prepared according to Jager et al. (1996b). The smoke water solutions significantly improved seed germination and rate (Moreira

Pycnanthemum pilosum

et al. 2010). This species is native to the Mediterranean region, but is now grown all over the world.

Salvia apiana **Jeps.**

WHITE SAGE

White sage is native to the Californian chaparrals of the United States. Keeley (1987) first tested charate on the seeds of this species and observed no effect on germination, or inhibitory effects if the seeds were heat treated (70°C for 60 minutes). Keeley and Fotheringham (1998a) have since shown that germination increased in response to aerosol smoke treatments.

Salvia azurea **Michx. ex Lam.**

BLUE SAGE

Chou et al. (2012) tested smoke water, heat, and combinations of both heat and smoke on the seeds of this species. Smoke treatments comprised of soaking seeds for 20 hours in the commercially available Regen 2000˚ smoke water solution, at concentrations of 1:5, 1:10, or 1:100 (v/v). The seeds were then exposed to heat treatments of 50 and 80°C for 5 minutes. Germination percentage was significantly increased when the seeds were exposed to the smoke water dilution of 1:100 at 50°C. Higher concentrations of smoke water preparations (1:10 and 1:5) inhibited germination percentages when tested at both 50°C and 80°C. Mean germination time was significantly increased for seeds treated with the 1:10 and 1:5 concentrations of smoke water on its own and was similarly delayed when tested at 50 and 80°C.

Salvia coccinea **L. f.**

BLOOD SAGE

Four minutes of smoke treatment, followed by 4 weeks of cold vernalization, significantly increased germination in this species (Zavala and Schwilk 2009). In a similar study by Schwilk and Zavala (2012), where they exposed the seeds to aerosol smoke treatments of 4 or 8 minutes, heat treatments at 30 or 60 seconds at 100°C and wet, cold stratification for 1 month at 4°C, there was no significant effect on germination. There was a significant increase in germination from 12% to 85% when the seeds were exposed to 4 minutes of aerosol smoke after they were stratified in dry, cold environment at 4°C for 1 month and incubated at a relative humidity of 10%. There was 50% germination when they were incubated in aerosol smoke for 8 minutes. In the absence of stratification, the smoke treatment significantly increased germination from 20% to 53%. This species occurs throughout the Americas.

Salvia columbariae **Benth.**

CHIA

Germination of chia, a native of the Californian chaparrals of the United States, was significantly increased by treating the seeds with 5% charate (Keeley 1984;

Keeley and Fotheringham 1998a) or aerosol smoke (5 and 8 minutes exposure) (Fotheringham et al. 1995; Keeley and Fotheringham 1998a).

Salvia farinacea **Benth.**

MEALY-CUP SAGE

Four minutes of smoke treatment, followed by 4 weeks of cold vernalization, significantly increased germination of the seeds of this species (Zavala and Schwilk 2009). Aerosol smoke treatments of 8 minutes, when combined with heat treatments at 30 or 60 seconds at 100°C and wet, cold stratification for 1 month at 4°C, did not significantly improve germination, which was 80% in the control group (Schwilk and Zavala 2012). A significant increase in germination occurred if the seeds were stratified in dry, cold environment at 4°C for 1 month and maintained at a relative humidity of 10%. Germination increased from 35% to 50%. This was following 8 minutes of exposure to the smoke. In the absence of stratification, germination increased from 22% to 39%. This species is native to Mexico and parts of the United States.

Salvia iodantha **Fernald.**

MEXICAN FUCHSIA SAGE

A variety of different fire cues were tested on the seeds of this species, which were collected from a mixed forest located in a mountainous subtropical area of Mexico (Zuloaga-Aguilar 2011). These included heat shock (100:15 sand: water [w:w] substrate at 120°C for 5 minutes), soaking the seeds in smoke water for 3 hours (prepared by burning 150 g of mixed forest litter and bubbling the resultant smoke into 1.5 L of distilled water, and adjusting the pH to 5 using sodium hydroxide) or ash (1.5 g of fine ash was added to the agar plates used to germinate the seeds). Combinations of these treatments were also tested for their effects on seed germination. The smoke and ash treatments, when applied on their own, did not significantly improve germination for this species. The heat shock and the combinations heat shock/smoke water and heat shock/ash did significantly increase it.

Salvia lavanduloides **Humb., Bonpl. and Kunth**

LAVENDER LEAF SAGE

Like *S. iodantha* above, a variety of different fire cues were tested on the seeds of this species, which were also collected from a mixed forest located in a mountainous subtropical area of Mexico (see *S. iodantha* for details; Zuloaga-Aguilar et al. 2011). Four of the treatments significantly increased germination percentages: ash, heat shock, heat shock and ash combination, and the combined ash/heat shock and smoke water treatment.

Salvia leucophylla **Greene**

SAN LUIS PURPLE SAGE

Salvia leucophylla is native of the Californian chaparrals of the United States. Keeley and Fotheringham (1998a) reported that germination of this species increased in response to aerosol smoke. This species is considered rare and at risk of becoming endangered on a global scale (IUCN 2013).

Salvia mellifera **Greene**

BLACK SAGE

Germination of black sage, a native of the Californian chaparrals of the United States, was significantly increased by treating the seeds with 5% charate (Keeley 1984; Keeley and Fotheringham 1998a).

Salvia penstemonoides **Kunth & Bouché**

BIG RED SAGE

Aerosol smoke treatments of 4 or 8 minutes, heat treatments at 30 or 60 seconds in an oven at 100°C and wet, cold stratification for 1 month at 4°C, did not significantly improve germination of the seeds of this species (Schwilk and Zavala 2012). Big red sage is native to Texas, United States, and is considered rare.

Salvia reflexa **Hornem.**

MINT WEED

Like *S. azurea* above, Chou et al. (2012) tested the effects of smoke water, heat, and combinations of both on the seeds of this species. Smoke treatments comprised of soaking the seeds for 20 hours in the commercially available Regen 2000˚ smoke water solution, at concentrations of 1:5, 1:10, or 1:100 (v/v). The seeds were then heated to 50 or 80°C for 5 minutes. A combination of the 1:5 smoke water dilution and both the 50 and 80°C heat shock treatments significantly decreased germination percentage while the 1:10 dilution interacted with 50°C to significantly increase it. Mean germination time was not affected by any of the treatments or interactions.

Salvia thyrsiflora **Benth.**

SALVIA

A variety of different fire cues were tested on the seeds of this species, which were collected from a mixed forest located in a mountainous subtropical area of Mexico (Zuloaga-Aguilar et al. 2011). These included heat shock (100:15 sand: water [w:w] substrate at 120°C for 5 minutes), soaking the seeds in smoke water for 3 hours (prepared by burning 150 g of mixed forest litter and bubbling the resultant smoke into 1.5 L of distilled water, and adjusting the pH to 5 using sodium hydroxide) or ash (1.5 g of fine ash was added to the agar plates used to germinate the seeds). Combinations of these treatments were also tested for their effects on seed germination. Treatment of seeds with smoke water, as well as the combination of smoke water, ash, and heat treatments, all resulted in significantly lower germination percentages when compared with the controls.

Salvia verbenaca **L.**

WILD SAGE

Wild sage is an invasive plant species that occurs in the semiarid and arid zones of Australia. Soaking 11-year-old seeds in smoke water for 1, 3, or 5 hours significantly inhibited their germination (Scally and Florentine personal communication).

Satureja thymbra **L.**

THYME-LEAVED SAVORY

Seed germination of this woody shrub from the Marmaris region of southwestern Turkey was significantly improved when its seeds were treated for 24 hours with aqueous smoke preparations derived from various plants (Çatav et al. 2012). Its germination rate was unaffected.

Sideritis angustifolia **Lam.**

RABO DE GATO

The seeds of this species, collected from the Mediterranean Basin, were incubated for 24 hours in two smoke water solutions of 1:1 and 1:10 concentrations, prepared according to Jager et al. (1996b). Neither of the smoke water solutions had any effect on germination percentage or rate and did not affect seedling growth (Moreira et al. 2010).

Teucrium capitatum **L.**

CAT THYME GERMANDER

Like *Sideritis angustifolia* above, the seeds of this species were collected from the Mediterranean Basin and incubated for 24 hours in two smoke water solutions of 1:1 and 1:10 concentrations, to test the effects of smoke on germination (Moreira et al. 2010). Neither of the smoke water solutions had any effect on germination or seedling growth.

Teucrium ronnigeri **Sennen**

POLEO DORATO

Like *Sideritis angustifolia* and *T. capitatum* above, the seeds of this species were also collected from the Mediterranean Basin and incubated for 24 hours in two smoke water solutions of 1:1 and 1:10 concentrations (Moreira et al. 2010). Neither of the smoke water solutions had any effect on germination or seedling growth.

Thymus piperella **L.**

CATALAN PEBRELLA

Like *T. ronnigeri* above, the seeds of this species were also collected from the Mediterranean Basin and incubated for 24 hours in two smoke water solutions of 1:1 and 1:10 concentrations (Moreira et al. 2010). The smoke water solutions had no effect on germination, but did significantly improve seedling growth (Moreira et al. 2010).

Thymus vulgaris **L.**

COMMON THYME

Common thyme is native to the Mediterranean region, but is now grown all over the world. Seeds from the Mediterranean Basin were incubated for 24 hours in two smoke water solutions of 1:1 and 1:10 concentrations, which were prepared according to Jager et al. (1996b). The smoke water solutions significantly improved seed germination, rate, and seedling growth (Moreira et al. 2010).

LAURACEAE

Cryptocarya alba **(Mol.) Looser**

PUEMO

Thirty minutes exposure to cool aerosol smoke significantly inhibited germination in this woody species of the Mediterranean matorral of central Chile (Gómez-González et al. 2008).

LILIACEAE

Calochortus splendens **Douglas ex Benth.**

SPLENDID MARIPOSA LILY

Splendid mariposa lily is native to the chaparral ecosystems of California. Germination of its seeds was inhibited by charate, with 89% germination (no pretreatment) decreasing to 66% germination (charate treatment) (Keeley and Keeley 1987).

Xerophyllum tenax **(Pursh) Nutt.**

COMMON BEAR GRASS

Smoke water treatment of common bear grass seed, collected from a wetland habitat and having undergone 14 weeks of cold stratification, significantly improved germination in this species (Shebitz et al. 2009).

LOASACEAE

Mentzelia dispersa **S. Watson**

BUSHY BLAZINGSTAR

A combination of smoke treatment and heat significantly promoted germination in the seeds of this postfire annual (Keeley et al. 2005). The bushy blazingstar is native to California but occurs in other parts of North America.

Mentzelia micrantha **(Hook. & Arn.) Torr. & Gray**

SAN LUIS BLAZINGSTAR

Mentzelia micrantha is native to the chaparrals of California. Germination of this species was promoted by charred wood leachate (5%), as well as to 5 and 8 minutes exposure to aerosol smoke (Keeley and Fotheringham 1998a).

MALVACEAE

Alyogyne hakeifolia **(Giord.) Alef.**

AUSTRALIAN HIBISCUS

This species is widely distributed throughout the southwestern region of Western Australia. Smoke water treatments of seeds collected at Salmon Gums, Western Australia, had no effect on the germination of this species, even when used in combination with other treatments, such as heat and manual scarification (Baker et al. 2005).

Alyogyne huegelii **(Endl.) Fryxell**

NATIVE HIBISCUS

This species occurs in sandy coastal plains from Geraldton to Esperance, Western Australia. Smoke water treatments of seeds collected from near Esperance had no effect on the germination of this species (Baker et al. 2005).

Iliamna rivularis **(Douglas ex Hook.) Greene var.** rivularis

STREAMBANK WILD HOLLYHOCK

Streambank wild hollyhock seeds, collected from a tallgrass prairie in the Midwest of the United States, did not germinate in response to aerosol smoke treatments of 1, 10, or 60 minutes duration (Jefferson et al. 2007). This species was reported as *Iliamna remota* in Jefferson et al. (2007).

Malva neglecta **Wallr.**

COMMON MALLOW

Common mallow is native to Europe, but is a weed in arable, disturbed sites of similar climatic regions. Seventy three percent of seeds germinated when treated with 10% smoke water in comparison to 28% germination in the control (Adkins and Peters 2001). Daws et al. (2007) also found that karrikinolide (10^{-7} M) promoted germination.

Sida subspicata **F. Muell. ex Benth.**

SPIKED SIDA

The spiked sida is native to Australia. A 60 minute aerosol smoke treatment of soil samples containing the seeds of this species, collected across forest edges between subtropical rainforests and eucalypt forests in the Lamington National Park of Queensland, Australia, did not promote germination (Tang et al. 2003).

MOLLUGINACEAE

Pharnaceum elongatum **(DC.) Adamson**

This South African fynbos species will not germinate unless the seeds are first treated with smoke. Forty-seven percent of aerosol smoked seeds germinated in a study conducted by Brown et al. (2003).

MONTINIACEAE

Montinia caryophyllacea **Thunb.**

WILD CLOVE BUSH

Wild clove bush occurs in the Cape Floristic Region of South Africa. Aerosol smoke had no effect on seed germination (Brown et al. 2003).

MORACEAE

Ficus coronata **Spin.**

SANDPAPER FIG

The sandpaper fig is native to eastern Australia. A 60 minute aerosol smoke treatment of soil samples containing seeds of this species, collected in the Lamington National Park of Queensland, Australia, did not promote germination in this species (Tang et al. 2003). An unidentified species of *Ficus* reported in this study also failed to germinate following the smoke treatment.

MYOPORACEAE

Myoporum turbinatum **Chinnock**

SALT MYOPORUM

This endangered plant species occurs only in one known area of the southwestern region of Western Australia. Freshly collected seeds were exposed for 60 minutes to cool aerosol smoke, but did not significantly respond to the treatment (Roche et al. 1997b).

MYRTACEAE

Actinodium cunninghamii **Schauer**

ALBANY DAISY

Dixon et al. (1995) reported that 90 minutes of cold smoke treatment had no effect on seed germination of this Western Australian species.

Agonis linearifolia **(DC.) Schauer**

SWAMP PEPPERMINT

Swamp peppermint occurs in the coastal regions of the southwestern region of Western Australia. Ex situ aerosol smoke applications significantly increased germination in this species (Roche et al. 1997b).

Astartea fascicularis **(Labill.) DC.**

FALSE BAECKEA

False baeckea is native to the proteaceous heathlands and jarrah forests of the Southwest Botanical Province of Western Australia. Ex situ aerosol smoke applications significantly increased germination in this species (Roche et al. 1997b).

Baeckea brevifolia **(Rudge) D.C.**

Baeckia brevifolia grows in the heathlands of New South Wales, Australia. Germination increased significantly in response to smoke when seeds were heated to 50°C for 5 minutes or not heated at all (Thomas et al. 2007).

Baeckea camphorosmae **Endl.**

CAMPHOR MYRTLE

This myrtle is widespread throughout the jarrah forests, banksia woodlands and heathlands of southwestern Western Australia. Ex situ aerosol smoke applications significantly increased germination in the species (Roche et al. 1997b).

Baeckea diosmifolia **Rudge**

FRINGED BAECKEA

Baeckea diosmifolia grows in the heathlands of southeastern Australia. Thomas et al. (2003) reported that germination of this species was unitive in response to aerosol smoke and heat treatments. Either treatment, when applied on its own, had no effect on germination. A combination of heat shock of 50°C for 5 minutes and aerosol smoke application for 20 minutes produced the best germination (approximately 35%) in comparison to the control (approximately 15%).

Baeckea imbricata **(Gaertn.) Druce**

HEATH MYRTLE

Like *B. diosmifolia* above, heath myrtle occurs in the heathlands of southeastern Australia. Thomas et al. (2003) reported that germination of this species was also unitive in response to aerosol smoke and heat treatments. Thomas et al. (2007) investigated this response further. They found that the germination response to smoke and heat was dependent on the population from which the seeds were collected. The germination response of population 1 exhibited a significant interaction between heat and smoke, whereas the germination response of population 2 was positive to smoke and negative to heat. Smoke increased germination of *B. imbricata* population 1 in the absence of heat.

Baeckea linifolia **Rudge**

WEEPING BAECKEA

The seeds of this species, when treated with smoke alone, did not exhibit any significant increase in germination, but did respond positively to a combination of smoke and heat shock (Thomas et al. 2007).

Baeckea **sp.**

The germination of seeds of an unidentified species of *Baeckia*, occurring naturally in a banksia woodland 20 km south of Perth, Western Australia, was not significantly improved following sprays of concentrated smoke water (50 and 100 mL/m^2) (Lloyd et al. 2000).

Callistemon speciosus **(Sims) Sweet**

ALBANY BOTTLEBRUSH

This species occurs in the jarrah forests of Western Australia. Neither aerosol smoke nor smoke water treatments significantly promoted germination of its seeds (Norman et al. 2006).

Calytrix aurea **Lindl.**

FRINGE MYRTLE

Dixon et al. (1995) reported that 90 minutes of cold smoke treatment had no effect on seed germination of this Western Australian species.

Calytrix breviseta **var.** breviseta **Lindl.**

SWAMP STARFLOWER

This rare and endangered calytrix is restricted to the banksia woodlands of the Swan Coastal Plain in Western Australia. Ex situ aerosol smoke applications significantly increased the germination of this species (Roche et al. 1997b).

Calytrix depressa **(Turcz.) Benth.**

Calytrix depressa is widely distributed throughout the southwest region of Western Australia. Ex situ aerosol smoke applications significantly increased germination in this species (Roche et al. 1997b).

Calytrix flavescens **A. Cunn.**

SUMMER STARFLOWER

Freshly collected seeds of this Australian plant species were exposed for 60 minutes to cool aerosol smoke without any significant effect on seed germination (Roche et al. 1997b).

Calytrix fraserii **Cunn.**

PINK SUMMER CALYTRIX

This species is widely distributed throughout the southwest of Western Australia. Ex situ aerosol smoke applications significantly increased its germination (Roche et al. 1997b).

Calytrix tetragona **Labill.**

COMMON FRINGE-MYRTLE

Calytrix tetragona is widely distributed throughout the southern and eastern regions of Australia. An ex situ aerosol smoke treatment significantly increased germination in this species (Roche et al. 1997b).

Chamelaucium **sp.**

WAX PLANT

An unidentified species of wax plant from Western Australia's proteaceous heathland, banksia woodland, and jarrah forest communities was part of a study by Cochrane et al. (2002), who aimed to identify plants whose seeds respond to smoke and its products. Removal of the seed coat, as well as soaking the seeds in concentrated smoke water for 24 hours and adding gibberellic acid (25 mg/L) to the growth medium, significantly promoted greater germination compared to using any one treatment alone.

Darwinia acerosa **W. Fitzg.**

FINE-LEAVED DARWINIA

Darwinia acerosa also occurs in Western Australia's proteaceous heathland, banksia woodland, and jarrah forest communities. Removal of the seed coat, as well as soaking the seeds in concentrated smoke water for 24 hours and adding gibberellic acid (25 mg/L) to the growth medium, significantly promoted greater germination compared to using any one treatment alone (Cochrane et al. 2002). This species is considered vulnerable (IUCN 2013).

Darwinia carnea **C. A. Gardner**

MOGUMBER

Like *D. acerosa* and *D. carnea* above, this species also occurs in Western Australia's proteaceous heathland, banksia woodland, and jarrah forest communities. Removal of the seed coat, as well as soaking the seeds in concentrated smoke water for 24 hours and adding gibberellic acid (25 mg/L) to the growth medium, significantly promoted greater germination compared to using any one treatment alone (Cochrane et al. 2002). This species is considered endangered and likely to become extinct (IUCN 2013).

Darwinia chapmaniana **N. G. Marchant**

CHAPMAN'S BELL

This is another of the *Darwinia* species that occurs in Western Australia's proteaceous heathland, banksia woodland, and jarrah forest communities. Removal of the seed coat, as well as soaking the seeds in concentrated smoke water for 24 hours and adding gibberellic acid (25 mg/L) to the growth medium, significantly promoted greater germination compared to using any one treatment alone (Cochrane et al. 2002).

Darwinia ferricola **N. G. Marchant**

SCOTT RIVER DARWINIA

Scott river darwinia is a rare and endangered (IUCN 2013) and native to the jarrah forest and *Agonis flexuosa* woodlands of Western Australia. Removal of the seed coat, as well as soaking the seeds in concentrated smoke water for 24 hours and adding gibberellic acid (25 mg/L) to the growth medium, significantly promoted greater germination compared to using any one treatment alone (Cochrane et al. 2002).

Darwinia lejostyla **(Turcz.) Domin**

Darwinia lejostyla is a rare and endangered (IUCN 2013) species of the proteaceous heathlands of Western Australia. Removal of the seed coat, as well as soaking the seeds in concentrated smoke water for 24 hours and adding gibberellic acid (25 mg/L) to the growth medium, significantly promoted greater germination compared to using any one treatment alone (Cochrane et al. 2002).

Darwinia **sp.**

Jarrah forest and banksia woodland communities of southwestern Australia form the home of this rare and endangered *Darwinia* species. Removal of the seed coat, as well as soaking the seeds in concentrated smoke water for 24 hours and adding gibberellic acid (25 mg/L) to the growth medium, significantly promoted greater germination compared to using any one treatment alone (Cochrane et al. 2002).

Darwinia squarrosa **(Turcz.) Domin**

FRINGED MOUNTAIN BELL

Fringed mountain bell is a native of southwestern Western Australia's proteaceous heathlands. It is considered vulnerable (IUCN 2013). Removal of the seed coat, as well as soaking the seeds in concentrated smoke water for 24 hours and adding gibberellic acid (25 mg/L) to the growth medium, significantly promoted greater germination compared to using any one treatment alone (Cochrane et al. 2002).

Darwinia wittwerorum **N. G. Marchant & Keighery**

WITTWER'S MOUNTAIN BELL

Southwestern Western Australia's proteaceous heathland is the home to another *Darwinia*. According to IUCN (2013), this species is considerednvulnerable. Removal of the seed coat, as well as soaking the seeds in concentrated smoke water for 24 hours and adding gibberellic acid (25 mg/L) to the growth medium, significantly promoted greater germination compared to using any one treatment alone (Cochrane et al. 2002).

Eremaea pauciflora **(Endl.) Druce**

Eremaea pauciflora occurs throughout the southwestern region of Western Australia. Smoke water sprays, with concentrations of 50 and 100 mL/m^2, did not significantly promote germination (Lloyd et al. 2000).

Eucalyptus blakelyi **Maiden**

BLAKELY'S RED GUM

Blakely's red gum is widespread and abundant in the woodland communities of eastern Australia. Germination was significantly inhibited (40%) when its seeds were treated with a 10% smoke water solution (Clarke et al. 2000).

Eucalyptus calophylla **Lindl.**

RED GUM

A 1 L/m² application of undiluted smoke water to the soil surface of a rehabilitated bauxite mine in the southwest of Western Australia had no significant effect on the germination of seeds from the soil seed bank (Roche et al. 1997a).

Eucalyptus dalrympleana **ssp.** heptantha **L. A. Johnson.**

MOUNTAIN WHITE GUM

Mountain white gum occurs in the eastern states of Australia. Smoke water treatments of seeds collected from the grassy woodlands of the New England Tableland of New South Wales had no effect on its germination (Clarke et al. 2000).

Eucalyptus marginata **Sm.**

JARRAH

Jarrah is one of the dominant tree species in the southwestern regions of Western Australia. Roche et al. (1997a) showed that germination of *E. marginata* seeds was significantly increased in response to in situ aerosol smoke and smoke water treatments of the soil surface of undisturbed jarrah forest soils and disturbed mine site soils, respectively.

Eucalyptus melliodora **A. Cunn. ex Schauer**

YELLOW BOX

Yellow Box occurs in the eastern states of Australia. Smoke water treatments of seeds collected from the grassy woodlands of the New England Tableland of New South Wales had no effect on its germination (Clarke et al. 2000).

Eucalyptus pauciflora **Sieb. ex Spreng**

SNOW GUM

Smoke water treatments of seeds collected from the grassy woodlands of the New England Tableland of New South Wales had no effect on its germination (Clarke et al. 2000).

Eucalyptus viminalis **Labill.**

MANNA GUM

Like several eucalypt species above, manna gum occurs in several Australian states. Smoke water treatments of seeds collected from the grassy woodlands of the New England Tableland of New South Wales had no effect on its germination either (Clarke et al. 2000).

Eucalyptus youmanii **Blakely & McKie**

YOUMAN'S STRINGYBARK

Eucalyptus youmanii occurs in the Australian states of Queensland and New South Wales. Smoke water treatments of seeds collected from the grassy woodlands of the New England Tableland of New South Wales had no effect on its germination (Clarke et al. 2000).

Hypocalymma angustifolium **(Endl.) Schauer**

PINK-FLOWERED MYRTLE

Hypocalymma angustifolium is native to Western Australia's proteaceous heath-land, banksia woodland, and jarrah forest regions. Ex situ germination significantly improved when the soil seed bank was treated with aerosol smoke for 90 minutes (Dixon et al. 1995). In contrast, Norman et al. (2007) reported that germination was inhibited by a 60-minute aerosol smoke treatment. Roche et al. (1997a) reported that aerosol smoke treatments of broadcast seeds used in rehabilitated mine sites in the southwest of Western Australia had no effect on their germination.

Hypocalymma robustum **(Endl.) Lindl.**

SWAN RIVER MYRTLE

The Swan River myrtle is commonly found growing in the banksia woodlands and jarrah forests of the southwestern region of Western Australia. Norman et al. (2007) reported that seeds treated with a 1% smoke water solution prior to sowing them on bauxite mine rehabilitation areas significantly improved germination. In contrast, Roche et al. (1997a) reported that neither aerosol smoke nor smoke water promoted germination in this species.

Kunzea ambigua **(Sm.) Druce**

TICK BUSH

Tick bush occurs in open *Eucalyptus* woodlands and heath of southeastern Australia. Heat shock treatment of 50°C for 5 minutes with 20 minutes of aerosol smoke treatment increased germination to over 80% (Thomas et al. 2003). Aerosol smoke treatment of 10 minutes alone resulted in approximately 60% germination. In another study, soil samples from the Eden Burning Study Area, a dry sclerophyll forest in the Yalumba State Forest of New South Wales, Australia, were collected and air dried to test the effects of heat, smoke, and an interaction between the two cues on seeds from the seed bank. Samples exposed to heat treatment were incubated at 80°C for 60 minutes while those exposed to smoke were incubated in a room, where smoke was generated for 120 minutes. Only the smoke treatment significantly improved the germination of the seeds of this species (Penman et al. 2008). Neither heat nor the interaction between the two cues had any effect.

Kunzea capitata **Rchb.**

PINK KUNZEA

Pink kunzea occurs in open *Eucalyptus* woodlands and heath of coastal areas and adjacent ranges in eastern Australia. Thomas et al. (2003) demonstrated that this species had a synergistic response to smoke and heat treatments. Over 60% germination occurred when a combination of heat shock (50°C for 5 minutes) and aerosol smoke (exposure for 20 minutes) was applied to the seed, in comparison to the control, in which less than 10% of seeds germinated. Smoke alone improved germination, whereas heat shock alone had no effect.

Kunzea ericoides **(A. Rich.) Joy Thomps.**

WHITE TEA TREE

Soil samples from the Eden Burning Study Area, a dry sclerophyll forest in the Yalumba State Forest of New South Wales, Australia, were collected and air dried to test the effects of heat, smoke, and an interaction between the two cues on seeds from the seed bank. Samples exposed to heat treatment were incubated at 80°C for 60 minutes while those exposed to smoke were incubated in a room, where smoke was generated for 120 minutes. The heat treatment on its own induced an increase in germination for this species, but was only marginally significant (Penman et al. 2008). Neither the smoke treatment nor interaction between the two cues had any effect.

Kunzea capitata

Kunzea pauciflora **Schauer**

MT. MELVILLE KUNZEA

Germination of this *Kunzea*, a native of Western Australia's proteaceous heathland, was significantly improved when seeds were soaked in concentrated smoke water for 24 hours. This was compared to seeds that received no treatment (79% and 5%, respectively) (Cochrane et al. 2002). This species is vulnerable and likely to become endangered in the near future (IUCN 2013).

Leptospermum myrsinoides **Schltdl.**

HEATH TEATREE

Germination of heath teatree seed, collected from a *Eucalyptus baxteri* heathy-woodland in Victoria, Australia, significantly improved following a combined heat and smoke water treatment (Enright and Kintrup 2001).

Leptospermum polygalifolium **Salisb.**

TANTOON YELLOW TEATREE

This species occurs in the Australian states of Queensland and New South Wales. Smoke water treatments of seeds collected from the grassy woodlands of the New England Tableland of New South Wales had no effect on germination (Clarke et al. 2000).

Leptospermum scoparium **J. R. Forst. & G. Forst.**

NEW ZEALAND TEA TREE

Soil samples from the Eden Burning Study Area, a dry sclerophyll forest in the Yalumba State Forest of New South Wales, Australia, were collected and air dried to test the effects of heat, smoke, and an interaction between the two cues on seeds from the seed bank. Samples exposed to heat treatment were incubated at 80°C for 60 minutes while those exposed to smoke were incubated in a room, where smoke was generated for 120 minutes. None of the treatments had any effect on the germination of the seeds of this species (Penman et al. 2008).

Leptospermum spinescens **Endl.**

SPINY TEATREE

Spiny teatree is common throughout the Southwest Botanical Province of Western Australia. Ex situ aerosol smoke applications significantly increased germination in this species (Roche et al. 1997b).

Lophostemon confertus **(R. Br.) Peter G. Wilson & J. T. Waterh.**

QUEENSLAND BRUSH BOX

The Queensland brush box is native to eastern Australia, but occurs elsewhere as an exotic plant. A 60 minute aerosol smoke treatment of soil samples containing seeds of this species, collected across forest edges between subtropical rainforests and eucalypt forests in the Lamington National Park of Queensland, Australia, did not improve germination in the species (Tang et al. 2003).

Micromyrtus ciliata **(Smith) Druce**

FRINGED HEATH MYRTLE

The fringed heath myrtle ranges from dry sclerophyll forests to heath habitats in New South Wales, Victoria and South Australia. Germination of this species is promoted by aerosol smoke (Thomas et al. 2007).

Scholtzia involucrata **(Endl.) Druce**

SPIKED SCHOLTZIA

Scholtzia involucrata, a native to Western Australian proteaceous heathlands, banksia woodlands, and jarrah forests, germinated more readily when concentrated smoke water (50 mL/m²) was applied in situ (Lloyd et al. 2000). Germination of its seeds was not, however, promoted when they were soaked in 10% smoke water for 24 hours (Clarke and French 2005).

Scholtzia laxiflora **Benth.**

PINK SCHOLTZIA

Dixon et al. (1995) reported that 90 minutes of cold smoke treatment had no effect on seed germination of this Western Australian species. The same applies to a 60-minute treatment of cold smoke (Roche et al. 1997b).

Thryptomene saxicola **Schauer**

ROCK THYRPTOMENE

The rock thyrptomene occurs on granite outcrops in the southwestern region of Western Australia. Freshly collected seeds of this species were treated for 60 minutes with cool aerosol smoke, but did not significantly respond to the treatment (Roche et al. 1997b).

Verticordia albida **A. S. George**

WHITE FEATHERFLOWER

White featherflower is a rare and endangered (IUCN 2013) native of the proteaceous heathland and mixed eucalypt woodlands of the southwestern region of Western Australia. A combination of seed pretreatments achieved high germination. Removal of the seed coat, soaking the seeds in concentrated smoke water for 24 hours and adding gibberellic acid (25 mg/L) to the germination medium resulted in significantly increased germination (Cochrane et al. 2002).

Verticordia attenuata **A. S. George**

Verticordia attenuata is rare and endangered and native to banksia woodland and jarrah forest communities of Western Australia's southwestern region. A combination of seed pretreatments achieved greater germination. Like *V. albida* above, removal of the seed coat, soaking the seeds in concentrated smoke water for 24 hours, and adding gibberellic acid (25 mg/L) to the germination medium significantly improved germination (Cochrane et al. 2002).

Verticordia aurea **A. S. George**

Verticordia aurea is native to the banksia woodlands and proteaceous scrub-heaths on the sandplains of coastal southwestern Western Australia. Ex situ aerosol smoke applications significantly increased the germination of this species (Roche et al. 1997b). This species is rare and at risk (IUCN 2013).

Verticordia bifimbriata **A. S. George**

Verticordia bifimbriata is rare and endangered and native to southwestern Western Australia's banksia woodland and jarrah forest communities. Like *V. albida* and *V. attenuata* above, a combination of seed pretreatments achieved greater germination. This comprised of removal of the seed coat, soaking the seeds in concentrated smoke water for 24 hours, and adding gibberellic acid (25 mg/L) to the germination medium (Cochrane et al. 2002).

Verticordia carinata **Turcz.**

STIRLING RANGE FEATHERFLOWER

The vulnerable (IUCN 2013) Stirling range featherflower is native to Western Australia's proteaceous heathland, banksia woodland, and jarrah forest. As for several previously mentioned *Verticordia* species, the combination of seed coat removal, soaking the seeds in concentrated smoke water for 24 hours and adding gibberellic acid (25 mg/L) to the germination medium significantly improved germination in this species (Cochrane et al. 2002).

Verticordia chrysantha **Endl.**

YELLOW FEATHERFLOWER

Yellow featherflower is commonly found in the mixed eucalypt woodlands of southwestern Western Australia. Seed germination was significantly increased if the seeds were stored in soil for 12 months and then treated with aerosol smoke for 60 minutes prior to being sown (Roche et al. 1997b).

Verticordia comosa **A. S. George**

This species of *Verticordia* also occurs in the proteaceous heathland and mixed eucalypt woodlands of the southwestern region of Western Australia. Removing the seed coat, followed by soaking the seeds in a concentrated smoke water solution for 24 hours and then adding gibberellic acid (25 mg/L) to the germination medium, significantly promoted germination in this species (Cochrane et al. 2002).

Verticordia dasystylis **spp. oestopoia A. S. George**

This is another of the *Verticordia* that occurs in the proteaceous heathland and mixed eucalypt woodlands of the southwestern region of Western Australia. According to Cochrane et al. (2002), a significant increase in germination was achieved by removal of the seed coat, soaking the seeds in smoke water for 24 hours, and germinating them on a medium to which gibberellic acid (25 mg/L) had been added.

Verticordia densiflora **Lindl.**

COMPACTED FEATHERFLOWER

Verticordia densiflora is native to proteaceous heathland, banksia woodland, and jarrah forest regions in Western Australia's Southwest Botanical Province. Ex situ germination significantly improved following a 90-minute treatment of its seeds (Dixon et al. 1995; see also Roche et al. 1997b). Germination also increased significantly when seeds were treated with karrikinolide (10 ppb), although the mean germination percentages were only 11% (Flematti et al. 2004).

Verticordia endlicheriana **var.** angustifolia **A. S. George**

This rare and endangered species occurs on granite outcrops in the jarrah forest region of Western Australia. Cochrane et al. (2002) achieved 24% germination by removing the seed coat and soaking the seeds in concentrated smoke water for 24 hours.

Verticordia eriocephala **A. S. George**

COMMON CAULIFLOWER

Yellow featherflower is commonly found in the mixed eucalypt woodlands of southwestern Western Australia. Ex situ aerosol smoke treatment of the seeds significantly increased germination in this species (Roche et al. 1997b).

Verticordia fimbrilepis **ssp.** australis **A. S. George**

SHY FEATHERFLOWER

This is another vulnerable (IUCN 2013) *Verticordia* species found growing on the granite outcrops in Western Australia's jarrah forest communities. Cochrane et al. (2002) used a combination of seed pretreatments to improve germination. Treatments included removing the seed coat, soaking the seeds in concentrated smoke water for 24 hours, and adding gibberellic acid (10 mg/L) to the germination medium.

Verticordia fimbrilepis **ssp.** fimbrilepis **Turcz.**

SHY FEATHERFLOWER

This species, which occurs in the proteaceous heathland, banksia woodland, and jarrah forest regions of Western Australia's Southwest Botanical Province, has been classified as endangered and likely to become extinct in the near future. Soaking the seeds in concentrated smoke water for 24 hours after removing the seed coat and adding gibberellic acid (10 mg/L) to the germination medium significantly improved germination (Cochrane et al. 2002).

Verticordia harveyii **Benth.**

AUTUMN FEATHERFLOWER

This rare and endangered (IUCN 2013) native to southwestern Western Australia's proteaceous heathlands, banksia woodlands, and jarrah forests exhibited significantly increased germination when soaking seeds in a concentrated smoke water solution for 24 hours (Cochrane et al. 2002). As part of the treatment, the seed coat was removed and gibberellic acid (25 mg/L) was added to the germination medium.

Verticordia helichrysantha **Benth.**

BARRENS FEATHERFLOWER

Barrens featherflower is considered vulnerable by the IUCN (2013). It is a native of the proteaceous heathlands of southwestern Western Australia. Germination of 63% was achieved by removing the seed coat and soaking the seeds in smoke water for 24 hours (Cochrane et al. 2002).

Verticordia huegelii **Endl.**

VARIEGATED FEATHERFLOWER

Variegated featherflower is native to the jarrah forest and banksia woodland regions of Western Australia. Ex situ aerosol smoke treatment of the seed significantly increased germination in this species (Roche et al. 1997b).

Verticordia hughanii **F. Muell.**

HUGHAN'S FEATHERFLOWER

Hughan's featherflower is considered endangered (IUCN 2013). It is a native of the proteaceous heathland and mixed eucalypt woodlands of the southwestern region of Western Australia. Removal of the seed coat, soaking the seeds in concentrated smoke water for 24 hours, and adding gibberellic acid (25 mg/L) to the germination medium resulted in significantly increased germination (Cochrane et al. 2002).

Verticordia plumosa **A. S. George var.** pleiobotrya

PLUMED FEATHERFLOWER

This rare and endangered (IUCN 2013) species can be found growing in the banksia woodlands and jarrah forests of the southwestern region of Western Australia. Cochrane et al. (2002) removed the seed coat and soaked the seeds for 24 hours in concentrated smoke water and achieved a final germination of 69%.

Verticordia plumosa **var.** vassensis **A. S. George**

PLUMED FEATHERFLOWER

Plumed featherflower is a rare small shrub which grows in the jarrah-marri forest, banksia and *Agonis flexuosa* woodlands and the proteaceous heathlands in the southwestern region of Western Australia. A combination of seed pretreatments achieved high germination. Removal of the seed coat, soaking the seeds in concentrated smoke water for 24 hours, and adding gibberellic acid (25 mg/L) to the germination medium resulted in significantly increased germination (Cochrane et al. 2002).

Verticordia spicata **A. S. George ssp.** squamosa

SPIKED FEATHERFLOWER

Spiked featherflower is native to the proteaceous heathlands of Western Australia. Cochrane et al. (2002) achieved 86% final germination of this rare and endangered shrub using a combination of seed pretreatments, which included soaking the seeds in smoke water for 24 hours.

Verticordia staminosa **A. S George ssp. cylindracea var.** cylindracea

WONGAN FEATHERFLOWER

This rare and endangered species occurs on the granite outcrops of the mixed eucalypt woodlands of Western Australia. According to Cochrane et al. (2002), a significant increase in germination was achieved by removal of the seed coat, soaking the seeds in smoke water for 24 hours and germinating them on a medium to which gibberellic acid (25 mg/L) had been added.

Verticordia staminosa **ssp. cylindracea A. S. George var.** erecta

WONGAN FEATHERFLOWER

This rare and endangered species also occurs on the granite outcrops of the mixed eucalypt woodlands of Western Australia. Cochrane et al. (2002) used a combination of seed pretreatments to improve germination. Treatments included removing the seed coat, soaking the seeds in concentrated smoke water for 24 hours and adding gibberellic acid (10 mg/L) to the germination medium.

Verticordia staminosa **C. A. Gardner & A. S. George ssp.** staminosa

WONGAN FEATHERFLOWER

This species grows on the granite outcrops in the proteaceous heathlands of south-western Western Australia. All of the seeds of this rare and endangered (IUCN 2013) species germinated when a combination of seed pretreatments was used, which included soaking the seeds in smoke water for 24 hours (Cochrane et al. 2002).

OLEACEAE

Fraxinus ornus **L.**

FLOWERING ASH

Crosti et al. (2006) reported that exposure to cool aerosol smoke for 60 minutes significantly decreased germination in this member of the Mediterranean plant community.

ONAGRACEAE

Camissonia californica **(Nutt. ex Torr. & Gray) Raven**

CALIFORNIA PRIMROSE

California primrose seed, which is native to the Californian chaparral communities of the United States, germinated only when seeds were treated with charred wood leachate (5% dilution) (Keeley and Keeley 1987; Keeley and Fotheringham 1998a) or aerosol smoke (5, 8, or 10 minutes exposure) (Keeley and Fotheringham

1998a). Germination also significantly increased in response to karrikinolide treatments of 10 ppb (Flematti et al. 2004).

Chamerion angustifolium **(L.) Holub**

FIREWEED

The seeds of this species, typically found in the ponderosa pine (*Pinus ponderosa*) forests of northern Arizona did not significantly respond to concentrated aqueous smoke treatment (Wright's Brand, Roseland, New Jersey) (Abella and Springer 2009).

Clarkia epilobioides **(Nutt. ex Torr. & Gray) A. Nelson & J. F. Macbr.**

WHITE CLARKIA

White clarkia is native to the chaparrals of northern Mexico, California, and Arizona. Germination significantly increased in response to a charate treatment of 0.5 g per Petri dish (Keeley and Keeley 1987).

Clarkia purpurea **(Curtis) A. Nelson & J. F. Macbr.**

PURPLE CLARKIA

This native of the chaparral ecosystems of the western United States and northern Mexico also responded positively to charate application (as described for *C. epilobioides*) (Keeley and Keeley 1987). Purple clarkia is considered rare and at risk of becoming endangered (IUCN 2013).

Epilobium angustifolium **L.**

FIREWEED

The effects of aerosol smoke, heat, darkness, cold stratification, and combinations of smoke with each of the three other treatments on seed germination were examined in this study (Tsuyuzaki and Miyoshi 2009). Smoke was produced by burning Timothy hay (*Phleum pratense*), which was pumped through a 3.5 m cooling tube into a smoke chamber for approximately 5 minutes. The seeds were exposed to the smoke for 60 minutes. Those seeds exposed also to heat were incubated at 75°C for 25 minutes. The cold stratification process took 1 month, during which the seeds remained in an incubator set to 4°C. Where the dark treatment was concerned, the seeds were maintained in total darkness for the entire germination period. The seeds of this species were collected over two different years (2004 and 2005) and were tested separately. While the actual seed germination percentages differed, their overall responses to the treatments were similar. Germination in both seed batches was reduced to 0% when the seeds of both years were treated with aerosol smoke. This was compared to 58% and 81% germination for the 2004 and 2005 control groups, respectively. Germination was significantly inhibited when the seeds were incubated in the dark. The 2004 seed batch resulted in 42% germination, while the 2005 batch achieved 59% germination. None of the other treatments had any effect on this species.

Epilobium cilliatum **Raf. ssp.** glandulosum **(Lehm.) Hoch & Raven**

FRINGED WILLOWHERB

Germination of the fringed willowherb, a native of the western and northern United States, Canada, and northern Mexico, was inhibited by aerosol smoke treatments (Jefferson et al. 2007). No treatment, 1 minute of aerosol smoke or 10 minutes of aerosol smoke had no effect on germination. However, 60 minutes exposure to aerosol smoke significantly decreased the number of germinants by 38%.

Fuchsia encliandra **Steund. ssp.** encliandra

FUCHSIA

A variety of different fire cues were tested on the seeds of this species, which were collected from a mixed forest located in a mountainous subtropical region of Mexico (Zuloaga-Aguilar et al. 2011). These included heat shock (100:15 sand: water (w:w) substrate at 120°C for 5 minutes), soaking the seeds in smoke water for 3 hours (prepared by burning 150 g of mixed forest litter and bubbling the resultant smoke into 1.5 L of distilled water, and adjusting the pH to 5 using sodium hydroxide) or ash (1.5 g of fine ash was added to the agar plates used to germinate the seeds). Combinations of these treatments were also tested for their effects on seed germination. The smoke water and ash treatments significantly promoted germination of the seeds of this species. Heat shock and heat shock combined with ash and/or smoke water treatments resulted in significantly lower germination percentages compared to the controls.

Oenothera biennis **L.**

COMMON EVENING PRIMROSE

The effects of aerosol smoke, heat, darkness, cold stratification, and combinations of smoke with each of the three other treatments on seed germination were examined in this study (Tsuyuzaki and Miyoshi 2009; see *Epilobium angustifolium* above for details of the tests performed). The dark treatment inhibited germination (89%) of this species compared to the control (100%). The seeds of this species display physiological dormancy.

ORCHIDACEAE

Oberonia ensiformis **(Sm) Lindl.**

SWORD-LEAF OBERONIA

Seed germination of this epiphytic orchid from the Karnataka State of India's Western Ghat Forest was significantly promoted (85%) (Malabadi et al. 2012) when treated with smoke saturated water that was prepared according to Thomas and van Staden (1995) and Dixon et al. (1995). The smoke water similarly had a significant effect on other aspects of the plant's development.

OROBANCHACEAE

Cistanche phelypaea **(L.) Cout.**

DESERT CANDLE

Significant germination in the seeds of this Saudi Arabia native occurred following treatment with the main germination active compound in smoke, karrikinolide (10^{-7} M; Daws et al. 2008). This response was similar to those induced by the synthetic strigol analogue, GR24 (10^{-6} M), and more effective compared to Nijmegan-1 (10^{-7} M).

OXALIDACEAE

Oxalis corniculata **L.**

PROCUMBENT YELLOW-SORREL

The origins of this species are unknown, but it occurs in many places in the world. A 60 minute aerosol smoke treatment of soil samples containing seeds of this species, collected across forest edges between subtropical rainforests and eucalypt forests in the Lamington National Park of Queensland, Australia, did not significantly increase germination in this species (Tang et al. 2003).

Oxalis **sp.**

Soil samples from the Eden Burning Study Area, a dry sclerophyll forest in the Yalumba State Forest of New South Wales, Australia, were collected and air dried to test the effects of heat, smoke, and an interaction between the two cues on seeds from the seed bank. Samples exposed to heat treatment were incubated at 80°C for 60 minutes while those exposed to smoke were incubated in a room, where smoke was generated for 120 minutes. None of the treatments had any effect on the germination of the seeds of an unidentified species of *Oxalis* (Penman et al. 2008).

PAPAVERACEAE

Papaver californicum **Gray**

FIRE POPPY

Papaver californicum grows in the chaparral and oak woodlands of central western and southwestern California, USA. Charate, applied at 0.5 g per Petri dish, significantly increased germination by 89% in comparison to the control (0%) (Keeley and Keeley 1987).

Papaver rhoeas **L.**

CORN POPPY

Corn poppy is a weedy species that has a broad native distribution throughout northern Africa, Europe and Asia. Karrikinolide (10^{-7} M) significantly promoted germination in this species (Daws et al. 2007).

Romneya coulteri **Harv.**

MATILIJA POPPY

Romneya coulteri is a rare (IUCN 2013) native of the Californian chaparrals of the United States and of western Canada. This species will not germinate without pretreatment of either charred wood (Keeley and Keeley 1987; Keeley and Fotheringham 1998a), aerosol smoke (Fotheringham et al. 1995; Keeley and Fotheringham 1998a) or smoke water (Keeley and Fotheringham 1998a).

Romneya trichocalyx **Eastw.**

BRISTLY MATILIJA POPPY

Romneya trichocalyx is restricted to the coastal chaparrals of southern California and northern Mexico. Germination of this poppy will only occur if seeds are treated with charate (0.25 g of powdered charred wood on filter paper) (Keeley 1987).

PAPILIONACEAE

Kennedia coccinea **Vent.**

CORAL VINE

Coral vine is native to Western Australian proteaceous heathlands, jarrah forests, and banksia woodlands. Roche et al. (1997a) discovered that germination of its seeds increased in response to smoke water treatment of the soil seed bank. This was part of a study aimed at using smoke to promote germination of native species used in mine site rehabilitation. Erratum: This species belongs in the Fabaceae family.

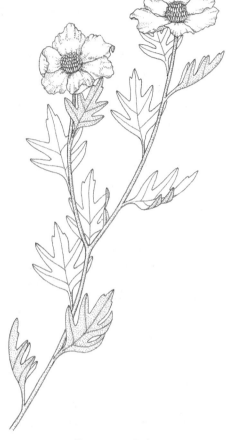

Romneya coulteri

PENAEACEAE

Endonema retzioides **Sond.**

Endonema retzioides is native to the Cape Floristic Region of South Africa. This species is classified as vulnerable by the IUCN (2013). Germination significantly increased in response to aerosol smoke treatment (Brown et al. 2003; Brown and Botha 2004).

Penaea **sp.**

This species is native to the Cape Floristic Region of South Africa. Germination significantly increased in response to aerosol smoke treatment (Brown et al. 2003).

PHILESIACEAE

Eustrephus latifolius **R. Br. ex Ker Gawl.**

WOMBAT BERRY

This species is native to Australia. A 60 minute aerosol smoke treatment of soil samples containing seeds of this species, collected across forest edges between subtropical rainforests and eucalypt forests in the Lamington National Park of Queensland, Australia, did not result in significantly increased germination (Tang et al. 2003).

PHYTOLACCACEAE

Phytolacca octandra **L.**

POKEWEED

Also known as inkweed, this species is a native of Tropical America, but has naturalized in Australia and elsewhere. A 60 minute aerosol smoke treatment of soil samples containing seeds of this species, collected from within the Lamington National Park of Queensland, Australia, did not significantly promote germination (Tang et al. 2003).

PINACEAE

Larix kaempferi **(Lam.) Carrière**

JAPANESE LARCH

The effects of aerosol smoke, heat, darkness, cold stratification, and combinations of smoke with each of the three other treatments on seed germination were examined in this study (Tsuyuzaki and Miyoshi 2009). Smoke was produced by burning Timothy hay (*Phleum pratense*), which was pumped through a 3.5 m cooling tube into a smoke chamber for approximately 5 minutes. The seeds were exposed to the smoke for 60 minutes. Those seeds exposed also to heat were incubated at 75°C for 25 minutes. The cold stratification process took 1 month, during which the seeds remained in an incubator set to 4°C. Where the dark treatment was concerned, the seeds were maintained in total darkness for the entire germination period. The seeds of this species were collected over two different years (2004 and 2005) and were tested separately. While the actual seed germination percentages differed, their overall responses to the treatments were similar. Germination was significantly inhibited by the combination of smoke and dark treatments. Specifically, the 2004 seed batch resulted in 2% germination following combined treatment compared to the control (9%). The 2005 seed batch resulted in 45% germination compared to the control (83%). None of the other treatments had any effect on this species, which is native to the eastern parts of North America and has naturalized in many other countries.

Pinus densiflora **Siebold & Zucc.**

JAPANESE RED PINE

The effects of aerosol smoke, heat, darkness, cold stratification, and combinations of smoke with each of the three other treatments on seed germination were examined in this study (Tsuyuzaki and Miyoshi 2009; see *Larix kaempferi* above for details about the tests performed). None of the treatments had any significant effect on the germination of this species.

Pinus douglasiana **Martínez**

DOUGLAS PINE

A variety of different fire cues were tested on the seeds of this species, which were collected from a mixed forest located in a mountainous subtropical area of Mexico (Zuloaga-Aguilar et al. 2011). These included heat shock (100:15 sand: water [w:w] substrate at 120°C for 5 minutes), soaking the seeds in smoke water for 3 hours (prepared by burning 150 g of mixed forest litter and bubbling the resultant smoke into 1.5 L of distilled water, and adjusting the pH to 5 using sodium hydroxide, or ash (1.5 g of fine ash was added to the agar plates used to germinate the seeds). Combinations of these treatments were also tested for their effects on seed germination. The smoke water, heat shock, and ash treatments, as well as the combination of heat shock and ash, and heat shock and smoke water all significantly improved germination percentages for this species.

PITTOSPORACEAE

Billardiera bicolor **(Putt.) E. M. Benn.**

PAINTED BILLARDIERA

Billardiera bicolor is common in Western Australia's proteaceous heathland, banksia woodlands, and jarrah forest regions. Ex situ germination significantly improved when the soil seed bank was treated with aerosol smoke for 90 minutes (Dixon et al. 1995).

Billardiera coeuleo-punctata **(Klotzsch) E. M. Benn.**

A 1 L/m² application of undiluted smoke water and aerosol smoke to the soil surface of a rehabilitated bauxite mine in the southwest region of Western Australia had no significant effect on the germination of seeds occurring in the soil seed bank (Roche et al. 1997a).

Billardiera procumbens **(Hook.) E. M. Benn.**

BILLARDIERA

Soil samples from the Eden Burning Study Area, a dry sclerophyll forest in the Yalumba State Forest of New South Wales, Australia, were collected and air dried to test the effects of heat, smoke, and an interaction between the two cues on seeds from the seed bank. Samples exposed to heat treatment were incubated at 80°C for 60 minutes while those exposed to smoke were incubated in a room, where smoke was generated for 120 minutes. Both heat and smoke on their own significantly improved germination of the seeds of this species while an interaction between the two cues had no effect (Penman et al. 2008).

Billardiera variifolia **DC.**

Blue billardiera occurs in the proteaceous heathland, banksia woodland, and jarrah forest communities of the southwestern region of Western Australia. Germination of *B. variifolia* was significantly improved in response to aerosol smoke treatment for 60 minutes prior to or after sowing (Roche et al. 1997a,b), but smoke water had no effect (Roche et al. 1997a). Norman et al. (2006) reported that neither aerosol smoke nor smoke water treatments significantly promoted germination in this species.

Bursaria spinosa **Cav.**

AUSTRALIAN BLACKTHORN

Bursaria spinosa has a widespread distribution in eastern Australia. Australian blackthorn germinated more readily when sown seeds had been treated with aerosol smoke for a period of 60 minutes (Roche et al. 1997b). *Bursaria spinosa* ssp. *spinosa* germination was also promoted by smoke water (10% dilution) treatment (Clarke et al. 2000).

Cheiranthera preissiana **Putt.**

Cheiranthera preissiana occurs in the proteaceous heathland, banksia woodland, and jarrah forest communities of the southwestern region of Western Australia. Germination was promoted when sown seeds were exposed to aerosol smoke for 60 minutes (Roche et al. 1997b).

Marianthus bicolor **(Putt.) F. Muell.**

This species is commonly found growing throughout the Southwestern Botanical Province of Western Australia. Germination was increased to 18% when its seeds were treated with 1% liquid smoke (Regen 2000® smoke water solution) and then sown on a bauxite mine rehabilitation area (Norman et al. 2007).

Marianthus floribunda **Putt.**

This species often occurs in the jarrah forests of Western Australia. Neither aerosol smoke nor smoke water treatments significantly promoted germination in its seeds (Norman et al. 2006).

Pronaya fraseri **(Hook.) E. M. Benn.**

ELEGANT PRONAYA

Elegant pronaya often occurs in the jarrah (*Eucalyptus marginata*) forests of Western Australia. Neither aerosol smoke nor smoke water treatments significantly promoted germination in its seeds (Norman et al. 2006).

Sollya heterophylla **Lindl.**

AUSTRALIAN BLUEBELL

Australian bluebell is a rare endemic of the proteaceous heathlands of the southwestern region of Western Australia. Its seeds germinated more readily when exposed to aerosol smoke for 60 minutes (Roche et al. 1997b), but not when treated with smoke water (Roche et al. 1997a) or when broadcast seeds used in rehabilitated mine sites in the southwest of Western Australia were treated with aerosol smoke. Norman et al. (2006) reported that neither aerosol smoke nor smoke water treatments significantly promoted germination in its seeds.

POLEMONIACEAE

Allophyllum glutinosum **(Benth.) Arn. & V. E. Grant**

STICKY FALSE GILYFLOWER

This gilyflower is native to the chaparral communities of California. Keeley and Fotheringham (1998a) reported that a 5-minute treatment of seeds with aerosol smoke significantly increased germination. However, 8 minutes of exposure had no effect.

Gilia capitata **Sims**

GLOBE GILIA

Globe gilia is commonly found growing throughout western Canada, the western United States and northern Mexico. Seed germination was promoted by a charate application of 0.5 g of powdered charred wood per Petri dish. Germination increased from 31% (no pretreatment) to 80% in response to charate (Keeley and Keeley 1987).

Saltugilia australis **(H. Mason & A. D. Grant) L. A. Johnson.**

SPLENDID GILIA

Saltugilia australis is native to the chaparral of southern California, and northern Mexico. Germination in this species was significantly promoted by a charate application (as described for *Gilia capitata* above) (Keeley and Keeley 1987).

POLYGALACEAE

Comesperma calymega **Labill.**

BLUE-SPIKE MILKWORT

Aerosol smoke treatments of broadcast seeds in rehabilitated mine sites in the southwest of Western Australia had no effect on their germination (Roche et al. 1997a).

Polygala lewtonii **Small**

LEWTON'S MILKWORT

Exposure to aerosol smoke for 5 minutes significantly promoted seed germination in this upland Florida plant species (Lindon and Menges 2008).

Comesperma virgatum **Labill.**

MILKWORT

Like *C. calymega*, aerosol smoke treatments of broadcast seeds in rehabilitated mine sites in the southwest of Western Australia did not significantly improve germination for this species (Roche et al. 1997a).

POLYGONACEAE

Fallopia convolvulus **(L.) Á. Löve**

BLACK BINDWEED

This invasive species is native to northern Africa, Europe, and Asia. Germination was significantly increased when seeds were treated with 20% smoke water (Adkins and Peters 2001).

Polygonella robusta **(Small) G. L. Nesom & V. M. Bates**

LARGEFLOWER JOINTWEED

Exposure to aerosol smoke for either 10 or 30 minutes inhibited seed germination in this upland Florida plant species (Lindon and Menges 2008).

Polygonum aviculare **L.**

KNOTWEED

Knotweed is a global weed in disturbed areas. Its origins are unknown. Smoke water (10–20% dilution) promoted germination, although it remained low (5% final germination) (Adkins and Peters 2001).

Polygonum longisetum **Bruijn**

CHANG LONG LIAO (CHINESE)

The effects of aerosol smoke, heat, darkness, cold stratification, and combinations of smoke with each of the three other treatments on seed germination were examined in this study (Tsuyuzaki and Miyoshi 2009). Smoke was produced by burning Timothy hay (*Phleum pratense*), which was pumped through a 3.5 m cooling tube into a smoke chamber for approximately 5 minutes. The seeds were exposed to the smoke for 60 minutes. Those seeds exposed also to heat were incubated at 75°C for 25 minutes. The cold stratification process took 1 month, during which the seeds remained in an incubator set to 4°C. Where the dark treatment was concerned, the seeds were maintained in total darkness for the entire germination period. Seed germination was negligible in the control group (0%–0.5%), but was promoted by the smoke treatment (4% germination), cold stratification (56%). The combination of smoke and cold stratification treatment significantly improved germination with 83% germination. None of the other treatments had any effect. This species occurs in China and other parts of Asia.

Polygonum pensylvanicum **L.**

PENNSYLVANIA SMARTWEED

Polygonum pensylvanicum is widely distributed throughout the United States and Canada and is considered a weed in many parts of the world. Germination doubled when smoke water (10%) was applied to the seeds (Adkins and Peters 2001).

Polygonum persicaria **L.**

LADY'S-THUMB

This native to Europe is a weed in temperate regions of the world. Germination was promoted by smoke water treatment (Adkins and Peters 2001).

Polygonum sachalinense **F. Schmidt ex Maxim.**

GIANT KNOTWEED

The effects of aerosol smoke, heat, darkness, cold stratification, and combinations of smoke with each of the three other treatments on seed germination were examined in this study (Tsuyuzaki and Miyoshi 2009; see *Polygonum longisetum* above for the details about the tests performed). Seed germination for the control group was low at 13%, but was inhibited further when the seeds were incubated in the dark, with germination reduced to 4%. No other treatments had an effect on the germination of this species, which occurs in the United States and parts of Asia.

Pterostegia drymarioides **Fisch. & C. A. Mey.**

WOODLAND PTEROSTEGIA

Pterostegia drymarioides is native to the western United States and northern Mexico. Keeley and Keeley (1987) reported that charate (0.5 g per Petri dish) inhibited germination in this species.

Rumex acetosella **L.**

SHEEP'S SORREL

Franzese and Ghermandi (2011) researched the effects of smoke and other fire factors on different aged seeds of *R. acetosella*, ranging from 1 to 19 years old. The seeds were exposed to 10 minutes of aerosol smoke. While there were no clear trends following exposure to smoke, the germination of fresh seeds was improved with the treatment (68% for the control versus 90% for the smoke treatment). In another study, the effects of aerosol smoke, heat, darkness, cold stratification, and combinations of smoke with each of the three other treatments on seed germination were examined in this study (Tsuyuzaki and Miyoshi 2009; see *Polygonum longisetum* above for details about the tests performed). The heat treatment promoted germination (40%) in this species compared to the control (27%). The dark and cold stratification treatments, as well as the combined smoke and dark, and smoke and cold stratification treatments significantly inhibited germination (18%, 9%, 11%, and 25%, respectively).

Rumex brownii **Campd.**

SWAMP DOCK

Swamp dock occurs throughout much of Australia. Smoke water treatments of seeds collected from the grassy woodlands of the New England Tableland of New South Wales had no effect on its germination (Clarke et al. 2000).

PRIMULACEAE

Anagallis arvensis **L.**

SCARLET PIMPERNEL

Scarlet pimpernel is native to Europe and has become a weed elsewhere in the world. Germination of this species was significantly promoted from 4 to 77 seedlings per m² when seeds were treated with aerosol smoke for 60 minutes (Read et al. 2000).

Coris monspeliensis **L.**

HIERBA DE LAS ÚLCERAS

The seeds of this species were incubated for 24 hours in two smoke water solutions of 1:1 and 1:10 concentrations, prepared according to Jager et al. (1996b). The smoke water solutions significantly improved seed germination percentage, rate, and seedling growth (Moreira et al. 2010). This species occurs in southern Europe and northern Africa.

PROTEACEAE

Adenanthos barbiger **Lindl.**

HAIRY JUGFLOWER

Hairy jugflower is endemic to the southwest region of Western Australia. A 1 L/m² application of undiluted smoke water to the soil surface of a rehabilitated bauxite mine in the southwest had no significant effect on the germination of seeds from the soil seed bank (Roche et al. 1997a).

Adenanthos ellipticus **A. S. George**

OVAL-LEAF ADENANTHOS

Oval-leaf adenanthos, classified as vulnerable (IUCN 2013), is native to the proteaceous heathlands of the southwestern region of Western Australia. A combination of nicking the seed coat to expose the endosperm, soaking the seeds in concentrated smoke water for 24 hours, and adding gibberellic acid (25 mg/L) to the germination medium was used to achieve a maximum germination of 11% (Cochrane et al. 2002).

Adenanthos pungens **Meisn. ssp.** pungens

SPIKY ADENANTHOS

This shrub can be found growing in the proteaceous heathlands and mixed eucalypt woodlands of the southwestern region of Western Australia. It is classified as vulnerable (IUCN 2013). A combination of treatments (as described for *A. ellipticus* above) was used to achieve 35% germination (Cochrane et al. 2002).

Aulax cancellata **(L.) Druce**

CHANNEL-LEAF FEATHERBUSH

Channel-leaf featherbush is native to the South African fynbos communities. Aerosol smoke treatments of 60 minutes duration to the soil seed bank (ex situ) significantly improved germination in this species (Brown et al. 1995). A later study revealed no significant effect following smoke treatment (Brown et al. 2003).

Aulax umbellata **R. Br.**

BROAD-LEAF FEATHERBUSH

Broad-leaf featherbush occurs in the Cape Floristic Region of South Africa. Aerosol smoke treatments had no effect on seed germination (Brown et al. 2003).

Banksia attenuata **R. Br.**

CANDLESTICK BANKSIA

The candlestick banksia is widely distributed throughout the southwestern region of Western Australia. Smoke water sprays, with concentrations of 50 and 100 mL/m^2, did not significantly promote germination of its seeds (Lloyd et al. 2000).

Banksia grandis **Willd.**

BULL BANKSIA

Bull banksia is native to southwestern Western Australia's proteaceous heathland, jarrah forest and banksia woodland regions. The germination of this tree significantly increased in response to in situ aerosol smoke treatments of the soil surface of a rehabilitated bauxite mine in the southwest of Western Australia (Roche et al. 1997a). A 1 L/m² application of undiluted smoke water to the soil surface had no effect on germination, however.

Banksia menziesii **R. Br.**

FIREWOOD BANKSIA

This species occurs only along the Swan Coastal Plain and Geraldton Sand Plains of Western Australia. Smoke water sprays, with concentrations of 50 and 100 mL/m², did not significantly promote germination of seeds collected in a banksia woodland 20 km south of Perth, Western Australia (Lloyd et al. 2000). Germination of its seeds was also not significantly promoted when they were soaked in 10% smoke water for 24 hours (Clarke and French 2005).

Conospermum huegelii **R. Br. ex Endl.**

Dixon et al. (1995) reported that 90 minutes of cold smoke treatment had no effect on seed germination of this Western Australian species.

Conospermum incurvum **Lindl.**

PLUME SMOKEBUSH

Plume smokebush is restricted to the banksia woodland communities of Western Australia. Dixon et al. (1995) reported that germination significantly increased when the soil in which it was sown was treated with aerosol smoke for 90 minutes. Roche et al. (1997b) observed maximum germination occurred when aerosol smoke treatments of 60 minutes duration were applied to seeds that had been in soil for 12 months.

Conospermum stoechadis **Endl.**

COMMON SMOKEBUSH

Dixon et al. (1995) reported that 90 minutes of cold smoke treatment had no effect on seed germination of this endemic shrub of Western Australia.

Conospermum triplinervium **R. Br.**

TREE SMOKEBUSH

Tree smokebush is native to Western Australia's proteaceous heathland, banksia woodland, and jarrah forest communities. Dixon et al. (1995) reported that germination of this species increased when sown seeds were treated with aerosol smoke for 60 minutes. Roche et al. (1997b) showed that maximum germination was achieved if seeds were stored in soil for 12 months and then treated with aerosol smoke for 60 minutes.

Dryandra nivea **(Labill.) R. Br.**

COUCH HONEYPOT

Couch honeypot often occurs in the jarrah forests of Western Australia. Neither aerosol smoke nor smoke water treatments significantly promoted germination of its seeds (Roche et al. 1997a; Norman et al. 2006).

Dryandra sessilis (Knight) Domin

PARROT BUSH

Parrot bush also often occurs in the jarrah forests of Western Australia. Like *D. nivea*, neither aerosol smoke nor smoke water treatments significantly promoted germination of its seeds (Roche et al. 1997a; Norman et al. 2006).

Grevillea althoferorum P. M.Olde & N. R. Marriott

This small rare and endangered shrub occurs in Western Australian proteaceous heathland, banksia woodland, and jarrah forest communities. Cochrane *et al.* (2002) used a combination of seed pretreatments to improve germination. These comprised of removing the seed coat, soaking the seeds in concentrated smoke water for 24 hours, and adding gibberellic acid (25 mg/L) to the germination medium. Fifty percent of the seeds germinated using this combination of treatments.

Grevillea buxifolia (Sm.) R. Br.

GREY SPIDER FLOWER

Grey spider flower is native to the dry sclerophyll woodlands or heathlands of New South Wales, Australia, and in its capital city of Sydney. Kenny (2000) reported that both heat and smoke were important germination cues for this species, with smoke being the more promotive of the two.

Grevillea buxifolia spp. buxifolia (Sm.) R. Br.

GREY SPIDER FLOWER

This subspecies of *Grevillea buxifolia* is also native to the dry sclerophyll woodlands or heathlands of New South Wales, Australia. Its seeds would not germinate unless they were first treated with 5 to 20 minutes of aerosol smoke or heat (80°C for 10 minutes). The best germination percentages (70%–80%) were the result of a combination of the two treatments (Morris 2000).

Grevillea calliantha R. O. Makinson & P. M. Olde

FOOTE'S GREVILLEA

Foote's grevillea is rare and endangered (IUCN 2013) and is a native of the Western Australian proteaceous heathland. One hundred percent germination was achieved by using a combination of treatments that involved nicking the seed coat to expose the endosperm, soaking the seeds in concentrated smoke water for 24 hours, and adding gibberellic acid (25 mg/L) to the germination medium (Cochrane et al. 2002).

Grevillea curviloba McGill. ssp. curviloba

Like *G. calliantha* above, this species is also a native of the Western Australian proteaceous heathland. Significant increases in germination were achieved by scratching the seed coat to expose the endosperm, soaking the seeds in concentrated smoke water for 24 hours, and adding gibberellic acid (25 mg/L) to the germination medium (Cochrane et al. 2002).

Grevillea curviloba **P. M. Olde & N. R. Marriott ssp.** incurva

NARROW CURVED-LEAVED GREVILLEA

The narrow curved-leaved grevillea occurs in the proteaceous heathlands of Western Australia. Exposing the endosperm, followed by soaking the seeds in concentrated smoke water for 24 hours and adding gibberellic acid (25 mg/L) to the germination medium, significantly improved germination in this species (Cochrane et al. 2002).

Grevillea diffusa **spp.** filipendula **McGill.**

CURVED-LEAF GREVILLEA

Curved-leaf grevillea occurs only in the dry sclerophyll woodlands of the coastal regions of central New South Wales. To promote germination, the seeds should be exposed to aerosol smoke for 5 to 20 minutes or, alternatively, to heat (80°C for 10 minutes) (Morris 2000). Combining the two treatments resulted in greater germination percentages.

Grevillea dryandroides **C. A. Gardner ssp.** dryandroides

PHALANX GREVILLEA

Significant increases in germination were achieved by scratching the seed coat to expose the endosperm, soaking the seeds in concentrated smoke water for 24 hours, and adding gibberellic acid (25 mg/L) to the germination medium (Cochrane et al. 2002). This species is considered endangered and likely to become extinct in the near future (IUCN 2013).

Grevillea elongata **P. M. Olde & N. R. Marriott**

IRONSTONE GREVILLEA

Grevillea elongata is a rare and endangered native of the Western Australian banksia woodland, mixed eucalypt woodland, and jarrah forest communities. Cochrane et al. (2002) used a combination of treatments, as described for *Grevillea dryandroides* ssp. *Dryandroides* above to achieve greater germination.

Grevillea eriostachya **Lindl.**

YELLOW FLAME GREVILLEA

Grevillea eriostachya seeds, were collected from Regens Ford and Cataby, Western Australia, and exposed to both smoke water (1:10 (v/v) dilution of smoke water; pH = 4.05, "Seed Starter," Kings Park and Botanic Garden, Perth, Western Australia) and karrikinolide (KAR_1; 1 μM) by Downes et al. (2010). The smoke water significantly increased germination while karrikinolide had no effect. The authors concluded that other chemicals in smoke may have promoted the germination.

Grevillea humifusa **P. M. Olde & N. R. Marriott**

SPREADING GREVILLEA

Spreading grevillea is a native of the proteaceous heathland, banksia woodland, and jarrah forest communities of the southwestern region of Western Australia. Significantly increased germination of this rare and endangered species was achieved using a combination of nicking the seed coat, smoke water and gibberellic acid (Cochrane et al. 2002).

Grevillea juniperina **R. Br.**

JUNIPER LEAF GREVILLEA

Juniper leaf grevillea has a widespread distribution in eastern New South Wales and southeastern Queensland, Australia. Morris (2000) achieved 60%–70% germination in comparison to less than 10% in the control, when its seeds were treated with a combination of dry heat (80°C for 10 minutes) and aerosol smoke (5–20 minutes).

Grevillea kenneallyi **McGill.**

High germination percentages of this rare and endangered species, which is native to Western Australia's proteaceous heathlands and mixed eucalypt woodlands, was achieved by exposing the endosperm, soaking the seeds in smoke water and adding gibberellic acid to the germination medium (Cochrane et al. 2002; see the *Grevillea* species above for more details).

Grevillea linearifolia **(Cav.) Druce**

WHITE SPIDER FLOWER

The distribution of *G. linearifolia* is mostly restricted to the dry sclerophyll forests of central eastern New South Wales, Australia. Morris (2000) achieved significantly greater germination when seeds of this species were treated with a combination of dry heat (80°C for 10 minutes) and aerosol smoke (5–20 minutes). There was a three-fold increase in germination when intact seeds were treated with smoke water, heat or a combination of the two, but was not significant possibly due to the low number of seeds and replications used.

Grevillea maccutcheonii **Keighery & Cranfield**

MCCUTCHEON'S GREVILLEA

McCutcheon's grevillea is a native of the proteaceous heathland, banksia woodland, and jarrah forest communities of the southwestern region of Western Australia. Significantly increased germination of this rare and endangered species was achieved using a combination of nicking the seed coat, smoke water, and gibberellic acid (Cochrane et al. 2002).

Grevillea maxwellii **McGill.**

MAXWELL'S GREVILLEA

This species occurs in Western Australia's proteaceous heathland, banksia woodland, and jarrah forest communities. Cochrane et al. (2002) used a combination of seed pretreatments to improve germination. These comprised of removing the seed coat, soaking the seeds in concentrated smoke water for 24 hours, and adding gibberellic acid (25 mg/L) to the germination medium. Fifty percent of the seeds germinated using this combination of treatments. This species is considered endangered and likely to become extinct in the near future (IUCN 2013).

Grevillea mucronulata **R. Br.**

GREEN SPIDER FLOWER

The distribution of *G. linearifolia* is mostly restricted to the dry sclerophyll forests of central eastern New South Wales, Australia. Germination ranging from 50% to 70% was achieved by treating seed with aerosol smoke for 5–20 minutes, or with aerosol smoke and heat (80°C for 10 minutes), or scarification, or a combination of the three treatments. The control resulted in 20% germination (Morris 2000).

Grevillea murex **McGill.**

This species of *Grevillea* occurs in the proteaceous heathlands of Western Australia. Exposing the endosperm, followed by soaking the seeds in concentrated smoke water for 24 hours and adding gibberellic acid (25 mg/L) to the germination medium, significantly improved germination in this species (Cochrane et al. 2002).

Grevillea pilulifera **(Lindl.) Druce**

WOOLLEY FLOWERED GREVILLEA

This grevillea is common to the hillsides and ridges of banksia woodland and jarrah-marri forest environments of Western Australia. Norman et al. (2007) reported that germination in this species was promoted by exposing its seeds to both aerosol (60 minutes) and liquid smoke (1%). Germination, however, was still low (7% and 3%, respectively) following treatment. Roche et al. (1997a) reported that treating broadcast seeds at rehabilitated mine sites in the southwest of Western Australia with aerosol smoke had no effect on the germination of this species.

Grevillea polybotrya **Meisn.**

Grevillea polybotrya is native to proteaceous heath-scrub and mixed eucalypt forests of the southwestern region of Western Australia. Germination of *G. polybotrya* increased when its seeds were exposed to aerosol smoke for 60 minutes (Roche et al. 1997b).

Grevillea pulchella **(R. Br.) Meissn.**

Roche et al. (1997a) reported that treating broadcast seeds at rehabilitated mine sites in the southwest of Western Australia with aerosol smoke had no effect on the germination of this species.

Grevillea quercifolia **R. Br.**

OAK-LEAF GREVILLEA

This species is native to the proteaceous heathland, banksia woodland, and jarrah forest communities of southwestern Western Australia. Germination of its seed was significantly improved when aerosol smoke was applied to them for 60 minutes (Roche et al. 1997b). When broadcast seeds in rehabilitated mine sites, also in the southwestern region of Western Australia, were treated with aerosol smoke, there was no significant improvement in germination (Roche et al. 1997a).

Grevillea scapigera **A. S. George**

CORRIGIN GREVILLEA

This rare and endangered (IUCN 2013) grevillea is native to the Western Australian proteaceous heathland and mixed eucalypt woodland communities. Maximum germination of 50% was achieved when its seeds were stored in soil for 12 months and then treated with aerosol smoke for 60 minutes (Roche et al. 1997b). Cochrane et al. (2002) achieved 55% germination using a combination of nicking the seed coat, smoke water, and gibberellic acid treatments (as described for other grevilleas above). Baker et al. (2005) reported that smoke water treatments of seed collected from a translocated population at Corrigin, Western Australia, had no significant effect on seed germination even when used in combination with other treatments.

Grevillea sericea **(Sm.) R. Br.**

SILKY GREVILLEA

Silky grevillea has a widespread distribution throughout eastern New South Wales and southeastern Queensland, Australia. Morris (2000) achieved 60%–70% germination, compared to less than 10% in the control, when its seeds were treated with a combination of dry heat (80°C for 10 minutes) and aerosol smoke (5–20 minutes). Kenny (2000) showed that both heat and smoke were important germination cues for this species, with smoke being the more promotive of the two.

Grevillea speciosa **(Knight) McGill.**

RED SPIDER FLOWER

The distribution of G. *speciosa* occurs in the dry sclerophyll forests of central eastern New South Wales, Australia, and in its capital city of Sydney. Significantly increased germination (50%–70%) was achieved when seeds were treated only with aerosol smoke for 5–20 minutes, or with aerosol smoke and heat (80°C for 10 minutes), or scarification, or a combination of the 3 treatments. The control resulted in 20% germination (Morris 2000). Kenny (2000) reported that both heat and smoke acted as germination cues for the species, with smoke being the more important of the two.

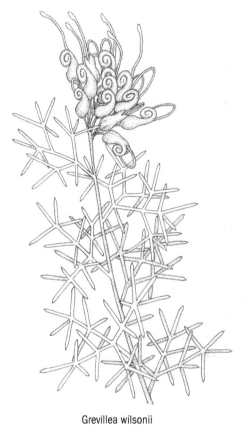

Grevillea wilsonii

Grevillea wilsonii **A. Cunn.**

WILSON'S GREVILLEA

Wilson's grevillea, which is native to southwestern Western Australia (specifically proteaceous heathlands, banksia woodlands, and jarrah forests), exhibited significantly increased germination percentages when treated with aerosol smoke (Dixon et al. 1995, Roche et al. 1997a,b, Morris et al. 2000). Norman et al. (2007) reported optimal germination (18%) when seeds were preimbibed with smoke water (1% dilution) and then incubated at an18/10°C and light/dark 9/13 hour regime.

Hakea amplexicaulis **R. Br.**

PRICKLY HAKEA

Roche et al. (1997a) reported that treating broadcast seed at rehabilitated mine sites in the southwest of Western Australia with aerosol smoke had no effect on the germination of this species.

Hakea cyclocarpa **Lindl.**

RAMSHORN

Like *C. amplexicaulis*, treating broadcast seeds at rehabilitated mine sites in the southwest of Western Australia with aerosol smoke had no effect on the germination of this species.

Hakea corymbosa **R. Br.**

CAULIFLOWER HAKEA

Cauliflower hakea is commonly found growing in the jarrah forests, mallee woodlands and proteaceous heathlands in the Southwest Botanical Province of Western Australia. Germination increased from 19% with no treatment to 75%, when sown seeds were exposed to 60 minutes of aerosol smoke (Roche et al. 1997b).

Hakea eriantha **R. Br.**

Hakea eriantha occurs in the wet sclerophyll forests of the coastal ranges of eastern Australia (Queensland, New South Wales, and Victoria). Germination of this species was inhibited by 54% when its seeds were germinated in Petri dishes containing 10% smoke water (Clarke et al. 2000).

Hakea lissocarpha **Lindl.**

HONEY BUSH

Aerosol smoke treatments of broadcast seeds in rehabilitated mine sites in the southwest of Western Australia had no effect on their germination (Roche et al. 1997a).

Hakea microcarpa **R. Br.**

SMALL FRUITED HAKEA

Smoke water treatments of seeds collected from the grassy woodlands of the New England Tableland of New South Wales had no effect on its germination (Clarke et al. 2000).

Hakea ruscifolia **Labill.**

CANDLE-SPIKE HAKEA

The candle-spike hakea commonly occurs in the southwest region of Western Australia. A 1 L/m² application of undiluted smoke water to the soil surface of a rehabilitated bauxite mine in the southwest of Western Australia had no significant effect on the germination of seeds taken from the soil seed bank (Roche et al. 1997a). Aerosol smoke treatments of broadcast seed similarly had no effect on germination.

Hakea stenocarpa **R. Br.**

NARROW-FRUITED HAKEA

Aerosol smoke treatments of broadcast seeds in rehabilitated mine sites in the southwest of Western Australia had no effect on their germination (Roche et al. 1997a).

Hakea undulata **R. Br.**

WAVY-LEAVED HAKEA

Hakea undulata is common in the banksia woodlands and jarrah forests of southwestern Western Australia. Roche et al. (1997a) reported a significant increase in germination in this species when the soil surface was treated with aerosol smoke, but smoked water had no effect. This study took place on a bauxite mine site undergoing rehabilitation.

Isopogon ceratophyllus **R. Br.**

HORNY CONEBUSH

Horny conebush is native to the coastal heathlands of western Victoria and eastern South Australia. Treatment with aerosol smoke of 60 minutes duration resulted in significantly greater germination (Roche et al. 1997b).

Leucadendron arcuatum **(Lam.) I. Williams**

RED-EDGED CONEBUSH

Red-edged conebush, a non-serotinous native South African fynbos species, exhibited a 1633% increase in germination in response to an aerosol smoke treatment of 30 minutes (Brown and Botha 2004).

Leucadendron conicum **(Lam.) I. Williams**

GARDEN-ROUTE CONEBUSH

Eighty percent germination was achieved for this South African fynbos species when seeds were treated with aerosol smoke for 30 minutes. This was compared to 59% for the control (Brown and Botha 2004).

Leucadendron coniferum **(L.) Meisn.**

DUNE CONEBUSH

Dune conebush is a native fynbos species of the Cape Floristic Region of South Africa. Brown et al. (2003) reported that aerosol smoke promoted its germination (see also Brown and Botha 2004).

Leucadendron daphnoides **(Thunb.) Meisn.**

DU TOIT'S KLOOF CONEBUSH

Aerosol smoke promoted the germination of *L. daphnoides* by 140%. A 1060% increase in germination was achieved when seeds were scarified and then treated with aerosol smoke for 30 minutes (Brown and Botha 2004). See also Brown et al. (2003). This species is native to South Africa and classified as vulnerable IUCN (2013).

Leucadendron gandogeri **Schinz ex Gand.**

BROAD-LEAVED CONEBUSH

Broad-leaved conebush occurs in the Cape Floristic Region of South Africa. Aerosol smoke had no effect on its seed germination (Brown et al. 2003).

Leucadendron laureolum **(Lam.) Fourc.**

This species of conebush also occurs in the Cape Floristic Region of South Africa. Aerosol smoke had no effect on its seed germination either (Brown et al. 2003).

Leucadendron linifolium **R. Br.**

LINE-LEAF CONEBUSH

Line-leaf conebush also occurs in the Cape Floristic Region of South Africa. Aerosol smoke had no effect on germination (Brown et al. 2003).

Leucadendron rubrum **Burm. f.**

SPINNINGTOP CONEBUSH

This conebush is a native fynbos species of the Cape Floristic Region of South Africa. Brown et al. (2003) reported that aerosol smoke significantly promoted germination in this species (see also Brown and Botha 2004).

Leucadendron salicifolium **(Salisb.) I. Williams**

COMMON STREAM CONEBUSH

This species occurs in the Cape Floristic Region of South Africa. Aerosol smoke had no effect on seed germination (Brown et al. 2003).

Leucadendron salignum **P. J. Bergius**

COMMON SUNSHINE CONEBUSH

This species is native to the Cape Floristic Region of South Africa. Brown et al. (2003) reported that aerosol smoke promoted its germination (see also Brown and Botha 2004).

Leucadendron sessile **R. Br.**

SUN CONEBUSH

This species occurs in the Cape Floristic Region of South Africa and elsewhere in southern Africa. Aerosol smoke had no effect on germination (Brown et al. 2003).

Leucadendron tinctum **I. Williams**

SPICY CONEBUSH

Spicy conebush is a native fynbos species of the Cape Floristic Region of South Africa. Brown et al. (2003) reported that aerosol smoke significantly improved seed germination in this species (see also Brown and Botha 2004).

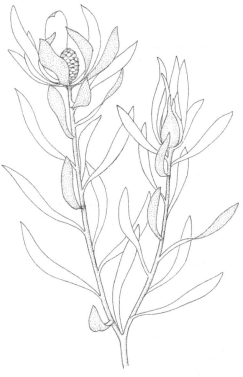

Leucadendron salignum

Leucadendron xanthoconus **K. Schum.**

SICKLE-LEAF CONEBUSH

Sickle-leaf conebush occurs in the Cape Floristic Region of South Africa and elsewhere in southern Africa. Aerosol smoke had no effect on germination (Brown et al. 2003).

Leucospermum cordifolium **(Salisb. ex Knight) Fourc.**

PINCUSHION

This species occurs in the Cape Floristic Region of South Africa and elsewhere in southern Africa. Aerosol smoke had no effect on seed germination either (Brown et al. 2003).

Leucospermum conocarpodendron **(L.) St. John.**

GREY-TREE PINCUSHION

Thirty minutes exposure to aerosol smoke did not significantly improve seed germination in this native fynbos species of the Cape Floristic Region of South Africa (Brown and Botha 2004).

Leucospermum glabrum **E. Phillips**

OUTENIQUA PINCUSHION

Like *L. conocarpodendron*, 30 minutes exposure to aerosol smoke did not significantly improve seed germination in this native fynbos species of the Cape Floristic Region of South Africa (Brown and Botha 2004).

Leucospermum praecox **Rourke**

LARGE TUFTED PINCUSHION

Like *L. conocarpodendron* and *L. glabrum*, 30 minutes exposure to aerosol smoke did not significantly improve seed germination in this native fynbos species of the Cape Floristic Region of South Africa (Brown and Botha 2004).

Leucospermum prostratum **(Thunb.) Stapf.**

YELLOW TRAILING PINCUSHION

Yellow trailing pincushion is a native fynbos species of the Cape Floristic Region of South Africa. Brown et al. (2003) reported that aerosol smoke significantly promoted its germination (see also Brown and Botha 2004).

Lomatia fraseri **R. Br.**

TREE LOMATIA

This species occurs in the eastern states of Australia. Smoke water treatments of seeds collected from the grassy woodlands of the New England Tableland of New South Wales had no effect on its germination (Clarke et al. 2000).

Lysinema ciliatum **R. Br.**

CURRY FLOWER

This species of *Lysinema* is native the proteaceous heathland, banksia woodland, and jarrah forest regions of the southwestern region of Western Australia. Ex situ germination was significantly improved when the soil seed bank was exposed to aerosol smoke for 90 minutes (Dixon et al. 1995).

Mimetes argenteus **Salisb. ex Knight**

SILVER PAGODA

Thirty minutes exposure to aerosol smoke did not significantly improve seed germination in this native fynbos species of the Cape Floristic Region of South Africa (Brown and Botha 2004).

Mimetes cucullatus **(L.) R. Br.**

RED PAGODA

Like *M. argenteus*, seed germination in this fynbos species of the Cape Floristic Region of South Africa was not significantly improved after 30 minutes of exposure to aerosol smoke (Brown and Botha 2004).

Persoonia elliptica **R. Br.**

SPREADING SNOTTYGOBBLE

This species occurs in the southwestern region of Western Australia. Freshly collected seeds of this species were treated for 60 minutes with cool aerosol smoke, but did not significantly respond to the treatment (Roche et al. 1997b).

Persoonia longifolia **R. Br.**

UPRIGHT SNOTTYGOBBLE

Dixon et al. (1995) reported that 90 minutes of cold smoke treatment had no effect on seed germination in this species, which is commonly found in the jarrah forests of southwestern Western Australia. Roche et al. (1997a,b) reported that neither aerosol smoke on broadcast seeds in rehabilitated mines nor smoke water had any effect on seed germination either.

Petrophile canescens **A. Cunn. ex R. Br.**

CONESTICKS

Conesticks is native to the heathlands and dry sclerophyll forests of coastal Queensland and New South Wales, Australia. Germination was significantly inhibited by smoke water application (as described for *Hakea eriantha*) (Clarke et al. 2000).

Petrophile drummondii **Meisn.**

This species occurs in the proteaceous heathland, banksia woodland, and jarrah forest regions of the southwestern region of Western Australia. Ex situ germination was significantly improved when the soil seed bank was exposed to aerosol smoke for 90 minutes (Dixon et al. 1995; see also Roche et al. 1997b).

Petrophile linearis **R. Br.**

PIXIE MOPS

Pixie mops is endemic to the southwestern region of Western Australia. Freshly collected seeds of this species were treated for 60 minutes with cool aerosol smoke, but did not respond to the treatment (Roche et al. 1997b).

Protea acuminata **Sims**

BLACKRIM SUGARBUSH

Seed germination was not significantly improved in this fynbos species of the Cape Floristic Region of South Africa after 30 minutes of exposure to aerosol smoke (Brown et al. 2003; Brown and Botha 2004).

Protea compacta **R. Br.**

BOT RIVER SUGARBUSH

Germination of Bot River sugarbush, a native fynbos species of the Cape Floristic Region, South Africa, was promoted by aerosol smoke when treated for 30 minutes (Brown et al. 1995; Brown et al. 2003; Brown and Botha 2004). Germination also significantly increased when seeds were soaked in smoke water (1:10 dilution) for 24 hours (Brown 1993a).

Protea cordata **Thunb.**

HEART LEAF SUGARBUSH

Heart leaf sugarbush is a native fynbos species of the Cape Floristic Region of South Africa. Brown et al. (2003) reported that aerosol smoke promoted its germination (see also Brown and Botha 2004).

Protea cynaroides **(L.) L.**

KING PROTEA

King protea occurs in the Cape Floristic Region of South Africa. Aerosol smoke had no effect on its seed germination either (Brown et al. 2003).

Protea eximia **(Knight) Fourc.**

DUCHESS PROTEA

The duchess protea also occurs in the Cape Floristic Region of South Africa. Aerosol smoke had no effect on germination (Brown et al. 2003).

Protea longifolia **Andrews**

LONGLEAF SUGARBUSH

Thirty minutes exposure to aerosol smoke did not significantly improve seed germination in this native fynbos species common to the Cape Floristic Region of South Africa (Brown et al. 2003; Brown and Botha 2004).

Protea magnifica **Andrews**

QUEEN PROTEA

The queen protea also occurs in the Cape Floristic Region of South Africa and elsewhere in the world. Aerosol smoke had no effect on seed germination (Brown et al. 2003).

Protea punctata **Meisn.**

WATER SUGARBUSH

Thirty minutes exposure to aerosol smoke did not significantly promote seed germination in this native fynbos species common to the Cape Floristic Region of South Africa (Brown et al. 2003; Brown and Botha 2004).

Protea repens **(L.) L.**

COMMON SUGARBUSH

Aerosol smoke treatments of 30 minutes duration did not significantly improve seed germination in this native fynbos species of the Cape Floristic Region of South Africa (Brown et al. 2003; Brown and Botha 2004).

Serruria florida **(Thunb.) Salisb. ex Knight**

BLUSHING BRIDE

Germination of blushing bride, a native to South Africa, was promoted when aerosol smoke was applied to the seeds for 60 minutes (Brown et al. 1995), but the smoke had no significant effect on germination when applied for 30 minutes (Brown et al. 2003; Brown and Botha 2004).

Serruria phylicoides **(P. J. Bergius) R. Br.**

Serruria phylicoides seed, which are native to the South African fynbos, responded favorably to smoke water treatment. Germination increased significantly when treated with smoke water (Brown 1993a). Aerosol smoke has since been shown to also promote germination in this species (Brown et al. 1995; Brown et al. 2003).

Serruria villosa **(Lam.) R. Br.**

GOLDEN SPIDERHEAD

This species is native to the Cape Floristic Region of South Africa. Brown et al. (2003) reported that aerosol smoke promoted its germination (see also Brown and Botha 2004).

Sphenotoma capitatum **(R. Br.) Lindl.**

PAPER HEATH

Sphenotoma capitatum is a native of Western Australia's proteaceous heathland, banksia woodland, and jarrah forest regions. Ex situ germination significantly improved when the soil seed bank was exposed to aerosol smoke for 90 minutes (Dixon et al. 1995).

Stirlingia latifolia **(R. Br.) Steud.**

BLUEBOY

Blueboy is native to proteaceous heathlands, banksia woodlands, and jarrah forests of the southwest of Western Australia. Dixon et al. (1995) reported that germination of this species increased significantly when its seeds were treated with aerosol smoke for 60 minutes. Roche et al. (1997b) showed that maximum germination could be achieved if the seeds were first stored in soil for 12 months and then treated with aerosol smoke (see also Tieu et al. 2001b). Smoke water sprays, with concentrations of 50 and 100 mL/m², did not, however, promote germination (Lloyd et al. 2000).

Synaphea acutiloba **Meissn.**

GRANITE SYNAPHEA

Dixon et al. (1995) reported that 90 minutes of cold smoke treatment had no effect on seed germination for this Western Australian species.

Synaphea petiolaris **R. Br.**

SYNAPHEA

Dixon et al. (1995) reported that 90 minutes of cold smoke treatment had no effect on seed germination forthis Western Australian species. Roche et al. (1997a) reported similar findings for aerosol smoke treatments of broadcast seeds on a rehabilitated mine site in the southwest of Western Australia.

PYROLACEAE

Pyrola incarnata **(DC.) Freyn**

The effects of aerosol smoke, heat, darkness, cold stratification, and combinations of smoke with each of the three other treatments on seed germination were examined in this study (Tsuyuzaki and Miyoshi 2008). Smoke was produced by burning Timothy hay (*Phleum pratense*), which was pumped through a 3.5 m cooling tube into a smoke chamber for approximately 5 minutes. The seeds were exposed to the smoke for 60 minutes. Those seeds exposed also to heat were incubated at 75°C for 25 minutes. The cold stratification process took 1 month, during which the seeds remained in an incubator set to 4°C. Where the dark treatment was concerned, the seeds were maintained in total darkness for the entire germination period. None of the seeds, including those in the control group, germinated following any of the treatments. The seeds of this species, which occurs in China and other parts of Asia, display morphological dormancy.

RANUNCULACEAE

Clematis flammula **L.**

FRAGRANT VIRGIN'S BOWER

Crosti et al. (2006) reported that exposure to 60 minutes of cool aerosol smoke significantly promoted germination in this species, which commonly occurs in fire-prone habitats of southern Europe and northern Africa.

Clematis glycinoides **DC.**

FOREST CLEMATIS

Forest clematis is native to eastern Australia. Tang et al. (2003) reported that germination of this vine was promoted in response to aerosol smoke treatment of 60 minutes duration. Clarke et al. (2000), in contrast, reported that smoke water had no significant effect.

Clematis lasiantha **Nutt.**

PIPESTEMS

Pipestems is common in the chaparral and open woodlands of California and northern Mexico. Charate significantly increased germination of this vine (Keeley 1987).

Clematis pubescens **Endl.**

COMMON CLEMATIS

Roche et al. (1997a,b) reported that the germination of this native of Western Australia's proteaceous heath-scrub communities increased significantly in response to 60 minutes of exposure to aerosol smoke prior to or after sowing. In contrast, Norman et al. (2007) showed that aerosol smoke (60 minutes) inhibited germination.

Clematis vitalba **L.**

OLD MAN'S BEARD

Crosti et al. (2006) reported that exposure to 60 minutes of cool aerosol smoke significantly promoted germination in this species. Old man's beard has an holarctic distribution and is commonly found in mid-European shrubberies and mountains.

RHAMNACEAE

Ceanothus americanus **L.**

NEW JERSEY TEA

New Jersey tea commonly occurs throughout the central and eastern United States and eastern Canada. Ex situ germination trials revealed that aerosol smoke promoted the germination of this species (Jefferson et al. 2007).

Ceanothus cuneatus **(Hook.) Nutt.**

BUCKBRUSH

Buckbrush is widely distributed throughout the chaparral, coastal scrub, and woodland ecosystems of northern Mexico (Baja region), California and Oregon. Keeley (1987) reported that the best germination results, 58%, could be achieved by first treating the seeds with heat (100°C for 5 minutes) and then treating them with charate. This comprised of 0.25 g of powdered charred wood on filter paper incubated in dark conditions.

Ceanothus leucodermis **Greene**

WHITE BARK CALIFORNIA LILAC

This rare (IUCN 2013) white bark California lilac occurs in the chaparral, coastal scrub, and woodland communities of Mexico and California. Charate had no effect on germination under dark conditions, except when the seeds were heated to 70°C for 60 minutes. Germination was significantly decreased as a result of the charate. Heat treatments significantly promoted germination (Keeley 1987).

Ceanothus megacarpus **Nutt.**

BIG POD CEANOTHUS

This native of the chaparral and coastal scrub of southern California requires heat to germinate. Keeley (1987) reported that heating the seeds to 120°C for 5 minutes significantly improved germination from 11% to 80% using heat shock. Charate inhibited germination following heat treatment and had no effect when used on its own.

Ceanothus oliganthus **Nutt.**

HAIRY CEANOTHUS

Ceanothus oliganthus is native to the chaparral and coastal scrub of California. This species is classified as vulnerable IUCN (2013). Charate had no effect on seed germination unless seeds were first given a heat shock treatment of 100 °C of 5 minutes duration. Charate had an inhibitory effect on germination when used in combination with heat. Heat (100°C for 5 minutes) alone promoted germination (Keeley 1987).

Cryptandra arbutiflora **Fenzl**

WAXY CRYPTANDRA

This species often occurs in the jarrah forests of Western Australia. Neither aerosol smoke nor smoke water treatments significantly promoted germination of its seeds (Norman et al. 2006).

Phylica buxifolia **L.**

BOX-LEAF PHYLICA

Seed germination in this fynbos species of the Cape Floristic Region of South Africa was not significantly improved following 30 minutes of aerosol smoke treatment (Brown and Botha 2004).

Phylica ericoides **L.**

HEATH-LEAVED PHYLICA

Seed germination in this fynbos species of the Cape Floristic Region of South Africa was not significantly improved following 30 minutes of aerosol smoke treatment (Brown et al. 2003; Brown and Botha 2004).

Phylica pubescens **Alt.**

FEATHERHEAD

Seed germination in this fynbos species of the Cape Floristic Region of South Africa was not significantly improved following 30 minutes of aerosol smoke treatment (Brown et al. 2003; Brown and Botha 2004).

Rhamnus alaternus **L.**

ITALIAN BUCKTHORN

Crosti et al. (2006) reported that exposure to cool aerosol smoke for 60 minutes significantly promoted earlier emergence of seedlings in this Italian plant species.

Rhamnus crocea **Nutt.**

REDBERRY BUCKTHORN

Redberry buckthorn is widely distributed throughout the chaparral, coastal scrub, and woodland communities of California and Arizona and northern Mexico. Keeley (1987) reported that germination under dark conditions was promoted when charate was used or if its seeds had been heated to 100°C for 5 minutes. Charate alone had no effect on seed germination.

Siegfriedia darwinioides **C. A. Gardner**

This species is common in the proteaceous heath-scrub of the southwestern region of Western Australia. Seed germination was significantly improved by aerosol smoke applications of the soil in which the seeds had been sown (Dixon et al. 1995).

Spyridium globulosum **(Labill.) Benth.**

BASKET BUSH

Basket bush is a native of Western Australia's proteaceous heathland, banksia wood-land, and jarrah forest regions. Ex situ germination significantly improved when the soil seed bank was exposed to aerosol smoke for 90 minutes (Dixon et al. 1995).

Trevoa quinquenervia **Gill. & Hook.**

Thirty minutes exposure to cool aerosol smoke significantly promoted germina-tion in this woody species of the Mediterranean matorral of central Chile (Gómez-González *et al.* 2008).

Trymalium ledifolium **Fenzl**

FOREST SAGE

Forest sage is native to the mediterranean regions of Western Australia, espe-cially the proteaceous heathland, banksia woodland, and jarrah forest communi-ties. Roche et al. (1997a) investigated the effects of smoke water application to the soil seed bank of a bauxite mine site and discovered that *T. ledifolium* germina-

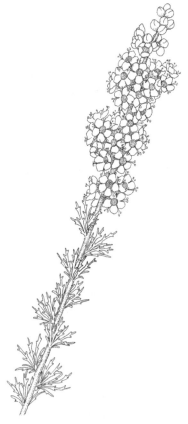

tion increased by 200% in response to that treat-ment. Interestingly, aerosol smoke treatments of broadcast seeds on a rehabilitated mine site in the southwest of Western Australia had no effect on their germination (Roche et al. 1997a). Norman et al. (2006) reported that neither aerosol smoke nor smoke water treatments significantly pro-moted germination in this species.

ROSACEAE

Acaena ovina **A. Cunn.**

SHEEP'S BUR

This species occurs in the eastern states of Aus-tralia. Smoke water treatments of seeds collected from the grassy woodlands of the New England Tableland of New South Wales had no effect on its germination (Clarke et al. 2000).

Adenostoma fasciculatum **Hook. & Arn.**

CHAMISE

Chamise is a shrub component of chaparral and woodland communities of northern Mexico, California, and Nevada. Keeley (1987) achieved

Adenostoma fasciculatum

significantly improved germination using charate

on seeds that had received a 60 minute heat treatment of 70°C. The final germination was 18%. A similar response was reported following smoke treatments combined with heat (Keeley et al. 2005).

Cliffortia ruscifolia **L.**

CLIMBER'S FRIEND

Climber's friend is often found growing on cliff faces or rocky slopes in the Cape Floristic Region of South Africa. Brown et al. (2003) revealed that aerosol smoke promoted germination in this species.

Kageneckia angustifolia **D. Don**

QUILLAY

Thirty minutes exposure to cool aerosol smoke significantly inhibited germination in this woody species of the Mediterranean matorral of central Chile (Gómez-González et al. 2008).

Kageneckia oblonga **Ruiz & Pav.**

BOLLÉN (SPANISH)

Like *K. angustifolia* above, 30 minutes exposure to cool aerosol smoke significantly inhibited germination in this woody species of the Mediterranean matorral of central Chile (Gómez-González et al. 2008).

Prunus grayana **Maxim.**

JAPANESE BIRD CHERRY

The effects of aerosol smoke, heat, darkness, cold stratification, and combinations of smoke with each of the three other treatments on seed germination were examined in this study (Tsuyuzaki and Miyoshi 2009). Smoke was produced by burning Timothy hay (*Phleum pratense*), which was pumped through a 3.5 m cooling tube into a smoke chamber for approximately 5 minutes. The seeds were exposed to the smoke for 60 minutes. Those seeds exposed also to heat were incubated at 75°C for 25 minutes. The cold stratification process took 1 month, during which the seeds remained in an incubator set to 4°C. Where the dark treatment was concerned, the seeds were maintained in total darkness for the entire germination period. Germination was low in the control group (1%) and none of the treatments had any effect on the germination of this species, which is native to temperate parts of Asia, including China and Japan.

Quillaja saponaria **Mol.**

SOAPBARK

Thirty minutes exposure to cool aerosol smoke also significantly inhibited germination in this woody species of the Mediterranean matorral of central Chile (Gómez-González et al. 2008).

Rubus moorei **F. Muell.**

SILKY BRAMBLE

Silky bramble is native to eastern Australia. A 60 minute aerosol smoke treatment of soil samples containing seeds of this and another unidentified species of *Rubus*, collected across forest edges between subtropical rainforests and eucalypt forests in the Lamington National Park of Queensland, Australia, did not germinate in response to the treatment (Tang et al. 2003).

Rubus parvifolius **L.**

NATIVE RASPBERRY

The effects of aerosol smoke, heat, darkness, cold stratification, and combinations of smoke with each of the three other treatments on seed germination were examined in this study (Tsuyuzaki and Miyoshi 2009; see *Prunus grayana* above for details about the tests performed). None of the seeds in the control group germinated or were affected by any of the treatments. This species is common in parts of East Asia, China, Japan, and Australia.

Sarcopoterium spinosum **(L.) Spach.**

THORNY BURNET

Seed germination of this woody shrub from the Marmaris region of southwestern Turkey was significantly improved when treated for 24 hours with aqueous smoke preparations derived from various plants (Çatav et al. 2012). Germination rate was not affected, however.

RUBIACEAE

Anthospermum spathulatum **Spreng.**

KOFFIEBOS

This species occurs in the Cape Floristic Region of South Africa and elsewhere in the world. Aerosol smoke had no effect on its seed germination (Brown et al. 2003).

Borreria radiata **DC.**

DAKA DAKE

Seeds collected from a Sudanian savanna-woodland in Burkina Faso, Africa, were treated with a variety of fire cues to determine their effects on seed germination (Dayamba et al. 2010). The seeds were soaked in smoke water (at concentrations of 100%, 75%, 50%, 25%, and 5% of the stock solution) for 24 hours. The smoke water stock solution was produced by burning a mixture of dominant native species from the Tiogo and Laba State forests of Sudan, and pumping the smoke through water for 10 hours. The seeds also underwent a heat shock treatment, during which they were incubated in an oven at 40, 80, 120, or 140°C for 2.5 minutes. Following these treatments, germination percentages and mean germination times were measured. None of the smoke water treatments had any effect on germination capacity. The 75% concentration of smoke water did, however, significantly increase mean germination time (more than double compared to the control). The heat shock treatments reduced germination capacity significantly with no germination occurring following some treatments.

Borreria scabra **K. Schum.**

BORRERIA

Like *B. radiata*, the seeds of this species were collected from a Sudanian savanna-woodland in Burkina Faso, Africa, and were also treated with a variety of fire cues to determine their effects on seed germination (see *B. radiata* for details; Dayamba et al. 2010). All of the smoke solutions significantly promoted germination capacity, but had no significant effect on the mean germination time. The heat shock treatments of 40 and 80°C also significantly increased germination capacity. Mean germination time doubled when the seeds were heated at 40, 80, and 120°C.

Galium angustifolium **Nutt. ex Gray**

BORREGO BEDSTRAW

This bedstraw is native to the chaparrals of California. Charate treatment of 0.5 g of powdered charred wood per Petri dish doubled the germination of this species (Keeley and Keeley 1987).

Galium aparine **L.**

STICKYWILLY

Stickywilly is native to Europe and Eurasia and a weed in the arable areas there. Germination significantly increased in response to smoke water treatment (Adkins and Peters 2001).

Galium migrans **Ehrend. & McGill.**

LOOSE BEDSTRAW

Loose bedstraw prefers moist sites in rocky crevices. This species is indigenous to the coastal regions of southern and eastern Australia. Germination was improved by exposing the soil seed bank to aerosol smoke for 60 minutes (Tang et al. 2003).

Galium **sp.**

Soil samples from the Eden Burning Study Area, a dry sclerophyll forest in the Yalumba State Forest of New South Wales, Australia, were collected and air dried to test the effects of heat, smoke, and an interaction between the two cues on seeds from the seed bank. Samples exposed to heat treatment were incubated at 80°C for 60 minutes while those exposed to smoke were incubated in a room, where smoke was generated for 120 minutes. The smoke treatment on its own induced an increase in germination of the seeds of an unidentified species of *Galium*, but was only marginally significant (Penman et al. 2008). Neither the heat treatment nor interaction between the two cues had any effect.

Oldenlandia galioides **(F. Muell.) F. Muell.**

This species is native to Australia and occurs in several states. A 60 minute aerosol smoke treatment of soil samples containing the seeds of this species, collected across forest edges between subtropical rainforests and eucalypt forests in the Lamington National Park of Queensland, Australia, did not germinate in response to the treatment (Tang et al. 2003).

Opercularia aspera **Gaertn.**

COARSE STINKWEED

Soil samples from the Eden Burning Study Area, a dry sclerophyll forest in the Yalumba State Forest of New South Wales, Australia, were collected and air dried to test the effects of heat, smoke, and an interaction between the two cues on seeds from the seed bank. Samples exposed to heat treatment were incubated at 80°C for 60 minutes while those exposed to smoke were incubated in a room, where smoke was generated for 120 minutes. Smoke treatment of the seeds of this species significantly improved their germination. The increase due to the heat treatment was only marginally significant (Penman et al. 2008). An interaction between the two cues had no effect.

Opercularia diphylla **Gaertn.**

STINKWEED

Stinkweed is widespread in sclerophyll forests, grasslands, and heathlands of New South Wales and Queensland, Australia. Germination was promoted when the soil seed bank was exposed to aerosol smoke for 60 minutes (Read et al. 2000).

Opercularia echinocephala **Benth.**

BRISTLY HEADED STINKWEED

This stinkweed is native to Western Australia, specifically the proteaceous heathland, jarrah forest and banksia woodland communities. Germination of *O. echinocephala* increased significantly when the soil surface was treated with aerosol smoke or smoke water (Roche et al. 1997a).

Opercularia hispidula **Endl.**

COARSE STINKWEED

Aerosol smoke treatments of broadcast seeds in rehabilitated mine sites in the southwest of Western Australia had no effect on their germination (Roche et al. 1997a).

Opercularia **sp.**

STINKWEED

Enright and Kintrup (2001) reported increased numbers of *Opercularia* species after treating the soil seed bank of a *Eucalyptus baxteri* heathy-woodland in Victoria, Australia with smoke water.

Opercularia varia **Hook. f.**

VARIABLE STINKWEED

Like *O. aspera* above, soil samples from the Eden Burning Study Area, a dry sclerophyll forest in the Yalumba State Forest of New South Wales, Australia, were collected and air dried to test the effects of heat, smoke, and an interaction between the two cues on seeds from the seed bank. Samples exposed to heat treatment were incubated at 80°C for 60 minutes while those exposed to smoke were incubated in a room, where smoke was generated for 120 minutes. Both heat and smoke treatments on their own significantly improved germination of the seeds of this species while an interaction between the two cues had no effect (Penman et al. 2008).

RUTACEAE

Agathosma betulina **(P. J. Bergius) Pillans**

ROUND LEAF BUCHU

Seed germination in this fynbos species of the Cape Floristic Region of South Africa was not significantly improved after 30 minutes exposure to aerosol smoke (Brown and Botha 2004). These researchers have noted, however, that the seeds tested may have been of poor viability or may require other special dormancy-breaking treatments.

Agathosma ovata **(Thunb.) Pillans**

FALSE BUCHU

Germination of false buchu, a native to the fynbos of South Africa, significantly increased in response to an aerosol smoke treatment of 30 minutes. Germination, however, was low regardless of the treatment with 10% germination when treated with smoke and 4% germination for the control (Brown and Botha 2004).

Agathosma tabularis **Sond.**

Seed germination in this species of the Cape Floristic Region of South Africa was not significantly improved after 30 minutes of aerosol smoke treatment (Brown et al. 2003; Brown and Botha 2004). These researchers have noted, however, that the seeds tested may have been of poor viability or may require special dormancy-breaking treatments.

Boronia fastigiata **Bartl.**

BUSHY BORONIA

Bushy boronia is native to the jarrah forest and proteaceous heath of the south-western region of Western Australia. The germination of this species responded positively to smoke water treatment, especially when applied to the soil seed bank (Roche et al. 1997a). In contrast, cold aerosol smoke treatment had no effect on germination (Dixon et al. 1995; Roche et al. 1997b). The same was reported for aerosol smoke treatments of broadcast seeds at a rehabilitated mine site in the southwest of Western Australia (Roche et al. 1997a).

Boronia megastigma **Nees ex Bartl.**

SCENTED BORONIA

Scented boronia occurs in the jarrah forest and proteaceous heaths of Western Australia. Roche et al. (1997b) were able to germinate the seeds of this species by applying aerosol smoke to sown seeds that had been stored in soil for 12 months.

Boronia spathulata **Lindl.**

BORONIA

This boronia occurs from Perth to Esperance in Western Australia. Freshly collected seeds of this species were exposed to 60 minutes of cool aerosol smoke, but did not significantly germinate in response to the treatment (Roche et al. 1997b).

Boronia tenuis **(Lindl.) Benth.**

BLUE BORONIA

This species is a rare (IUCN 2013), native to Western Australia's jarrah forest and banksia woodland communities. Blue boronia's germination was significantly promoted when sown seeds were exposed to aerosol smoke for 60 minutes (Roche et al. 1997b).

Boronia megastigma

Boronia viminea **Lindl. ssp.** opercularia

ANISEED BORONIA

Aniseed boronia has a broad distribution throughout the Southwest Botanical Province of Western Australia. Roche et al. (1997b) showed that germination increased slightly when using aerosol smoke as a pretreatment.

Coleonema album **E. Mey.**

CAPE MAY

Seed germination of this species of the Cape Floristic Region of South Africa was not significantly improved following 30 minutes of aerosol smoke treatment (Brown and Botha 2004). These researchers have noted, however, that the seeds tested may have been of poor viability or may require special dormancy-breaking treatments.

Correa reflexa **var.** cardinalis **(Labill.) Vent.**

NATIVE FUSCHIA

This species is endemic to Australia. Freshly collected seeds were treated for 60 minutes with cool aerosol smoke, but did not significantly respond to the treatment (Roche et al. 1997b).

Diosma acmaeophylla **Eckl. & Zeyh.**

RIBBOKBOEGOE (AFR.).

Seed germination of this species of the Cape Floristic Region of South Africa was not significantly improved following 30 minutes of aerosol smoke treatment (Brown and Botha 2004). These researchers have noted, however, that the seeds tested may have been of poor viability or may require special dormancy-breaking treatments.

Diospyros glabra **(L.) De Winter**

This species occurs in South Africa's Cape Floristic Region. An application of aerosol smoke did not improve seed germination in this species (Brown et al. 2003).

Diplolaena dampieri **Desf.**

SOUTHERN DIPLOLAENA

Diplolaena dampieri occurs in Western Australia's proteaceous heathlands. Germination of this species increased when seed that had already been sown were exposed to aerosol smoke for 60 minutes (Roche et al. 1997b).

Diplolaena grandiflora **Desf.**

WILD ROSE

Both smoke water and karrikinolide significantly improved germination in this species, which is commonly used in rehabilitation practices of disturbed areas of Western Australia (Commander et al. 2009a).

Eriostemon spicatus **A. Rich.**

PEPPER AND SALT

Pepper and salt can be found in the banksia woodlands of the Southwest Botanical Province of Western Australia. Dixon et al. (1995) showed that exposure of seeds to aerosol smoke for 90 minutes significantly increased germination. In contrast, Roche et al. (1997a) reported that aerosol smoke treatments of broadcast seeds in rehabilitated mine sites in the southwest of Western Australia had no effect on their germination.

Geleznowia verrucosa **Turcz.**

YELLOW BELLS

Geleznowia verrucosa is a native of Western Australia's proteaceous heathland, banksia woodland, and jarrah forest regions of the southwest. Ex situ germination significantly improved when the soil seed bank was exposed to aerosol smoke for 90 minutes (Dixon et al. 1995, see also Roche et al. 1997b).

Phebalium anceps **DC.**

BLISTER BUSH

Dixon et al. (1995) reported that 90 minutes of cold smoke treatment had no effect on seed germination of this Western Australian species.

Philotheca spicata **(A. Rich.) Paul G.Wilson**

PEPPER AND SALT

Pepper and salt is commonly found in banksia woodlands, jarrah forests, and proteaceous heathlands of southwestern Western Australia. Smoke water (1%) treatment produced a small but significant increase in the number of germinants when the seeds were sown on bauxite mine rehabilitation areas (Norman et al. 2007). Lloyd et al. (2000) reported that germination was not significantly promoted by an in situ application of concentrated smoke water (50 or 100 mL/m^2) to the soil seed bank in a banksia woodland 20 km south of Perth, Western Australia.

SANTALACEAE

Anthobolus foveolatus **F. Muell.**

Dixon et al. (1995) reported that 90 minutes of cold smoke treatment had no effect on seed germination for this Western Australian species.

Choretrum glomeratum **R. Br.**

BERRY BROOMBUSH

Dixon et al. (1995) reported that 90 minutes of cold smoke treatment had no effect on seed germination for this Western Australian species.

Exocarpus sparteus **R. Br.**

BROOM BALLART

Seed germination of broom ballart, a component of the coastal heathlands of Western Australia, significantly increased in response to 60 minutes exposure to aerosol smoke, but only when applied directly to seeds already sown in the ground (Roche et al. 1997b). Dixon et al. (1995) reported, in contrast, that 90 minutes of cold smoke treatment had no effect on seed germination.

Leptomeria cunninghamii **Miq.**

This species is found growing in the jarrah forests of Western Australia. Neither aerosol smoke nor smoke water treatments significantly promoted germination of its seeds (Norman et al. 2006).

SAPINDACEAE

Dodonaea viscosa **Jacq.**

FLORIDA HOPBUSH

This native of the United States occurs in many parts of the world. Smoke water treatments of seeds collected from the grassy woodlands of the New England Tableland of New South Wales had no effect on its germination (Clarke et al. 2000).

SAXIFRAGACEAE

Hydrangea paniculata **Raf.**

PEEGEE HYDRANGEA

The effects of aerosol smoke, heat, darkness, cold stratification, and combinations of smoke with each of the three other treatments on seed germination were examined in this study (Tsuyuzaki and Miyoshi 2008). Smoke was produced by burning Timothy hay (*Phleum pratense*), which was pumped through a 3.5 m cooling tube into a smoke chamber for approximately 5 minutes. The seeds were exposed to the smoke for 60 minutes. Those seeds exposed also to heat were incubated at 75°C for 25 minutes. The cold stratification process took 1 month, during which the seeds remained in an incubator set to 4°C. Where the dark treatment was concerned, the seeds were maintained in total darkness for the entire germination period. The cold stratification treatment resulted in 95% germination, while there was no germination when the seeds were incubated in the dark (0%–0.5%). When the smoke treatment was combined with heat, there was 40% germination. The smoke and cold stratification treatments, when combined, resulted in 87% germination. Seed germination of the control group was 50%. The seeds of this species displays physiological dormancy.

SCROPHULARIACEAE

Castilleja integra **A. Gray**

WHOLELEAF INDIAN PAINTBRUSH

The seeds of this species, typically found in the ponderosa pine (*Pinus ponderosa*) forests of northern Arizona, did not significantly respond to concentrated aqueous smoke treatments (Wright's Brand, Roseland, New Jersey) (Abella and Springer 2009).

Chenopodiopsis chenopodioides **(Diels) Hilliard.**

This native of the South African Cape Floristic Region germinated more readily when treated with aerosol smoke (Brown et al. 2003; Brown and Botha 2004).

Chenopodiopsis hirta **(L. f.) Hilliard.**

Like *C. chenopodioides* above, this native of the South African Cape Floristic Region also germinated more readily when treated with aerosol smoke (Brown et al. 2003; Brown and Botha 2004).

Collinsia parryi **Gray**

PARRY'S BLUE-EYED MARY

Germination of this native of the chaparrals of southern California was significantly improved by treatment with charate (Keeley and Keeley 1987). This species is classified this species as globally vulnerable (IUCN 2013).

Conopholis alpina **Liebm.**

MEXICAN CANCER-ROOT

Significant germination of the seeds of this native of Mexico occurred following treatment with the main germination compound in smoke, karrikinolide (10^{-6} M; Daws et al. 2008). This response was similar to those induced by the synthetic strigol analogue, GR24 (10^{-6} M), and more effective compared to Nijmegan-1 (10^{-7} M).

Digitalis obscura **L.**

SUNSET FOXGLOVE

The effects of two smoke water solutions on the seeds were tested by Moreira et al. (2010). The smoke water was prepared according to Jager et al. (1996b) and consisted of a 1:1 and 1:10 dilutions. The smoke water solutions had no effect on germination percentage or rate and did affect seedling growth. This species is common in the Mediterranean Basin and is native to Spain and parts of Africa.

Diplacus aurantiacus **Jeps. ssp.** aurantiacus

ORANGE BUSH MONKEY FLOWER

Orange bush monkey flower, previously known as *Mimulus aurantiacus*, is widely distributed from Oregon to

Diplacus aurantiacus ssp. aurantiacus

southern California and the California Baja of northern Mexico. Keeley (1987) showed that charate (0.25 g of powdered charred wood on filter paper) promoted seed germination under dark conditions.

Diplacus clevelandi (Brandegee) Greene

CLEVELANDS MONKEY FLOWER

Clevelands monkey flower was previously known as *Mimulus clevelandi*, and is a species classified as vulnerable (IUCN 2013). It is native to the Californian chaparral. Keeley and Fotheringham (1998a) reported that the germination of its seeds significantly increased in response to aerosol smoke treatment.

Dischisma capitatum (Thunb.) Choisy

WOOLLY-HEADED DISCHISMA

This native of the South African Cape Floristic Region germinated more readily when treated with aerosol smoke (Brown et al. 2003; Brown and Botha 2004).

Hebenstretia paarlensis Roessler.

This native of the South African Cape Floristic Region germinated more readily when treated with aerosol smoke (Brown et al. 2003; see also Brown and Botha 2004).

Keckiella cordifolia (Benth.) Straw

CLIMBING PENSTEMON

Climbing penstemon, which is native to southern California and northern Mexico, germinated readily without treatment. However, heat treatment at 100°C for 5 minutes significantly inhibited germination. Interestingly, charate promoted germination to levels equivalent to that of the control (no treatment) by counteracting the effects of the heat treatment (Keeley 1987).

Lathraea squamaria L.

TOOTHWORT

Germination in this species was significantly promoted following treatment with the active compound in smoke, karrikinolide (10^{-7} M; Daws et al. 2008). This response was similar to those induced by the synthetic strigol analogue, GR24 (10^{-6} M), and more effective than Nijmegan-1 (10^{-7} M).

Manulea cheiranthus (L.) L.

This native of the South African Cape Floristic Region germinated more readily when treated with aerosol smoke (Brown et al. 2003; see also Brown and Botha 2004).

Mimulus bolanderi **A. Gray**

BOLANDER'S MONKEYFLOWER

Keeley et al. (2005) reported that germination of fresh seeds of this Californian endemic can be promoted with smoke and heat, but the effects were not significant after a period of storage in soil.

Mimulus gracilipes **B. L. Rob**

SLENDERSTALK MONKEYFLOWER

Like *M. bolanderi* above, the germination of freshly collected seeds of this Californian species was significantly promoted following a smoke and heat treatment. The effects were, however, not significant after a period of storage in soil.

Nemesia lucida **Benth.**

This native of the South African Cape Floristic Region germinated more readily when treated with aerosol smoke (Brown et al. 2003; Brown and Botha 2004).

Nemesia strumosa **Benth.**

CAPEJEWELS

Seed germination of this species of the Cape Floristic Region of South Africa was not significantly improved following 30 minutes of aerosol smoke treatment (Brown and Botha 2004). These researchers noted, however, that the seeds tested may have been of poor viability or may require special dormancy-breaking treatments.

Nemesia versicolor **E. Mey. ex Benth.**

This native of the South African Cape Floristic Region germinated more readily when treated with aerosol smoke (Brown et al. 2003; Brown and Botha 2004).

Neogaerrhinum strictum **(Hook. & Arn.) Rothm.**

KELLOG SNAPDRAGON

Neogaerrhinum strictum is also known by its synonym, *Antirrhinum kelloggi*. This species is native to the Californian chaparral. Keeley and Fotheringham (1998a) reported that germination of its seeds significantly increased in response to treatment with aerosol smoke.

Orobanche aegyptiaca **Pers.**

EGYPTIAN BROOMRAPE

Orobanche aegyptiaca is a parasitic weed that is native to northern Africa, Asia, and eastern Europe. Germination was stimulated by smoke water (created by burning 4 g of filter paper and allowing absorption in 3 mL distilled water) dilutions of 1:5, 1:10, and 1:20. At high concentrations, smoke water was less effective (Bar Nun and Mayer 2005). Germination in the seeds of this species was also significantly promoted following treatment with the active compound in smoke, karrikinolide

(10^{-7} M; Daws et al. 2008). This response was similar, but lower, than those induced by the synthetic strigol analogues, GR24 (10^{-6} M) and Nijmegan-1 (10^{-7} M).

Orobanche caryophyllacea **Sm.**

BEDSTRAW BROOMRAPE

Significant germination of the seeds of this native of England occurred following treatment with karrikinolide (10^{-7} M; Daws et al. 2008). This response was similar to those induced by the synthetic strigol analogue, GR24 (10^{-6} M), and more effective compared to Nijmegan-1 (10^{-7} M).

Orobanche cernua **Loefl.**

NODDING BROOMRAPE

Seed germination of this species was significantly promoted following treatment with karrikinolide (10^{-7} M; Daws et al. 2008). This response was similar to those induced by the synthetic strigol analogues, GR24 (10^{-6} M) and Nijmegan-1 (10^{-7} M).

Orobanche corymbosa **(Rydb.) Ferris**

FLAT TOP BROOMRAPE

Significant germination of the seeds of this native of the United States occurred following treatment with karrikinolide (10^{-7} M; Daws et al. 2008). This response was similar to those induced by the synthetic strigol analogues, GR24 (10^{-6} M) and Nijmegan-1 (10^{-7} M).

Orobanche minor **Sm.**

HELLROOT

Germination in the seeds of this native of New Zealand was significantly promoted following treatment with karrikinolide (10^{-7} M; Daws et al. 2008). This response was similar, but lower, than those induced by the synthetic strigol analogue, GR24 (10^{-6} M), and greater than Nijmegan-1 (10^{-7} M).

Orobanche purpurea **Jacq.**

YARROW BROOMRAPE

Significant germination of the seeds of this native of England occurred following treatment with karrikinolide (10^{-7} M; Daws et al. 2008). This response was similar to those induced by the synthetic strigol analogues, GR24 (10^{-6} M) and Nijmegan-1 (10^{-7} M).

Orobanche ramosa **L.**

HEMP BROOMRAPE

Significant germination in the seeds of this South African native occurred following treatment with karrikinolide (10^{-7} M; Daws et al. 2008). This response was similar to those induced by the synthetic strigol analogue, Nijmegan-1 (10^{-7} M), and less effective compared to GR24 (10^{-6} M).

Orobanche rapum-genistae **Thuill.**

GREATER BROOMRAPE

Significant germination in the seeds of this native of Belgium occurred following treatment with karrikinolide (10^{-7} M; Daws et al. 2008). This response was similar to those induced by the synthetic strigol analogue, Nijmegan-1 (10^{-7} M), and less effective compared to GR24 (10^{-6} M).

Orobanche uniflora **L.**

ONE-FLOWERED BROOMRAPE

Germination in the seeds of this native of Canada was significantly promoted following treatment with karrikinolide (10^{-7} M; Daws et al. 2008). This response was similar to those induced by the synthetic strigol analogue, Nijmegan-1 (10^{-7} M), and less effective compared to GR24 (10^{-6} M).

Paulownia tomentosa **(Thunb.) Steud.**

EMPRESS TREE

Empress tree is native to eastern Asia, and is invasive in temperate regions of the world. Smoke water significantly promoted germination of its seeds when treated with red light. Smoke water did not induce germination under dark conditions. However, gibberellins could induce germination under dark conditions. Optimal concentrations of gibberellins, for inducing germination of seeds in the dark, were much lower when used in combination with smoke water (Todorović et al. 2005).

Penstemon barbatus **(Cav.) Roth**

BEARDLIP PENSTEMON

Penstemon barbatus commonly occurs in the fire-prone *Pinus ponderosa* forests of the southwestern and southcentral United States and northern Mexico. Abella (2006) tested smoke water (5%, 10%, or 20%), aerosol smoke (15 minutes), charred wood (5.5–5.6 g per pot) and charred wood leachate (10 g/100 mL) on the germination of this species. Smoke water significantly promoted germination (50%–60%) in comparison to the control (5%–20%), while the other three treatments had no effect. Soil type also influenced germination results. See also Abella and Springer (2009).

Penstemon centranthifolius **Benth.**

SCARLET BUGLER

Scarlet bugler is native to northern Mexico and California. Keeley and Keeley (1987) reported that the germination of this species was promoted by exposing its seeds to charate (0.5 g of powdered charred wood per Petri dish). Keeley and Fotheringham (1998a) showed that germination was also enhanced when seeds were treated with aerosol smoke for 5 or 8 minutes.

Penstemon cobaea **Nutt.**

COBAEA BEARDTONGUE

A significant improvement to germination occurred following an aerosol smoke treatment of 8 minutes, heat treatments at 30 or 60 seconds 100°C and wet, cold stratification for 1 month at 4°C (Schwilk and Zavala 2012). Germination percentages were not as high when the seeds were stratified for 1 month in a dry, cold environment at 4°C, and at a relative humidity of 10%, but were significant nevertheless.

Penstemon frutescens **Lamb.**

IWABUKURO

The effects of aerosol smoke, heat, darkness, cold stratification, and combinations of smoke with each of the three other treatments on seed germination were examined in this study (Tsuyuzaki and Miyoshi 2008). Smoke was produced by burning Timothy hay (*Phleum pratense*), which was pumped through a 3.5 m cooling tube into a smoke chamber for approximately 5 minutes. The seeds were exposed to the smoke for 60 minutes. Those seeds exposed also to heat were incubated at 75°C for 25 minutes. The cold stratification process took 1 month, during which the seeds remained in an incubator set to 4°C. Where the dark treatment was concerned, the seeds were maintained in total darkness for the entire germination period. The cold stratification treatment more than doubled the germination (52%) of this species when compared to the control group (24%). However, germination following dark treatment, as well as the combined smoke and heat and combined smoke and cold stratification treatment significantly inhibited germination (0%, 3%, and 9%, respectively). The seeds of this species display physiological dormancy. This species is common in parts of temperate Asia, especially on volcanoes.

Penstemon heterophyllus **Lindl.**

FOOTHILL PENSTEMON

Foothill penstemon is common throughout the chapparal communities of California. Exposing the seeds of this species seeds to charate (0.5 g of powdered charred wood per Petri dish) significantly promoted germination in them (Keeley and Keeley 1987).

Penstemon heterophyllus

Penstemon pachyphyllus **A. Gray ex Rybd.**

THICKLEAF BEARDTONGUE

The seeds of this species were collected from a frequently burned ponderosa pine forest (*Pinus ponderosa*) in northern Arizona and treated with concentrated aqueous smoke (Wright's Brand, Roseland, New Jersey). This significantly increased the percentage of emerging seedlings (Abella and Springer 2009).

Penstemon palmeri **A. Gray.**

PALMER'S PENSTEMON

Like *P. pachyphyllus* above, the seeds of this species were also collected from a ponderosa pine forest (*Pinus ponderosa*) in northern Arizona and treated with concentrated aqueous smoke (Wright's Brand, Roseland, New Jersey). This significantly increased germination (Abella and Springer 2009).

Penstemon rostriflorus **Kellogg**

BRIDGE PENSTEMON

Seeds collected from a ponderosa pine forest (*Pinus ponderosa*) in northern Arizona were treated with concentrated smoke water (Wright's Brand, Roseland, New Jersey), which significantly increased germination (Abella and Springer 2009).

Penstemon spectabilis **Thurb. ex Gray**

SHOWY PENSTEMON

Showy penstemon is native to northern Mexico and southern California. Germination of this species increased from 1% with no treatment to 61% when seeds were treated with charate (as described for *P. centranthifolius* above) (Keeley and Keeley 1987).

Penstemon virgatus **A. Gray.**

UPRIGHT BLUE BEARDTONGUE

Like *Penstemon* species above, the seeds of this species were also collected from a ponderosa pine forest (*Pinus ponderosa*) in northern Arizona and treated with concentrated aqueous smoke (Wright's Brand, Roseland, New Jersey). This significantly increased germination (Abella and Springer 2009).

Pseudoselago serrata **(P. J. Berg.) Hilliard**

BLOUAARBOSSIE

Seed germination of this species of the Cape Floristic Region of South Africa was not significantly improved after 30 minutes of exposure to aerosol smoke (Brown and Botha 2004). These researchers noted that the seed tested may have been of poor viability or may require special dormancy-breaking treatments.

Pseudoselago spuria **Hilliard**

Seed germination of this species of the Cape Floristic Region of South Africa was not significantly improved following 30 minutes of aerosol smoke treatment (Brown and Botha 2004). The viability of the seeds used in this study may have been low.

Sairocarpus coulterianus **(Benth. ex A. DC.) D. A. Sutton**

COULTER'S SNAPDRAGON

Charred wood leachate (5%) and aerosol smoke (5 and 8 minutes) seed treatments significantly increased germination of *S. coulterianus*, a native of the chaparral communities of California (Keeley and Fotheringham 1998a; see also Keeley et al. 1985; Keeley and Keeley 1987).

Sairocarpus nuttallianus **(Benth. ex A. DC.) D. A. Sutton**

VIOLET SNAPDRAGON

Sairocarpus nuttallianus is native to the Californian chaparrals of the United States. Keeley and Fotheringham (1998a) showed that germination of this species could be significantly increased in response to aerosol smoke.

Scrophularia californica **Cham. & Schltdl.**

CALIFORNIA FIGWORT

Scrophularia californica is native to the western United States and western Canada, and is often found growing in moist soils. Keeley and Keeley (1987) reported that germination was inhibited from 82% (no treatment) to 25% in response to charate (as described for *Penstemon centranthifolius*).

Selago **sp.**

An unidentified native *Selago* species of the South African Cape Floristic Region germinated more readily when treated with aerosol smoke (Brown et al. 2003; Brown and Botha 2004).

Striga hermonthica **Benth.**

PURPLE WITCHWEED

Germination in this species was significantly promoted following treatment with karrikinolide (10^{-7} M; Daws et al. 2008). This response was similar to those induced by the synthetic strigol analogue, GR24 (10^{-6} M), and more effective than Nijmegan-1 (10^{-7} M).

Veronica hederifolia **L.**

IVY-LEAF SPEEDWELL

This native of North America is a weed in parts of Australia. A commercially available smoke water solution, "Seed Starter," did not significantly improve seed germination (Adkins and Peters 2001).

Veronica notabilis **F. Muell. ex Benth.**

FOREST SPEEDWELL

This species is native to eastern Australia. A 60 minute aerosol smoke treatment of soil samples containing the seeds of this species, collected across forest edges between subtropical rainforests and eucalypt forests in the Lamington National Park of Queensland, Australia, did not germinate (Tang et al. 2003).

Veronica persica **Poir.**

BIRDEYE SPEEDWELL

Veronica persica has a wide distribution throughout Europe. Adkins and Peters (2001) showed that smoke water treatments (10% or 20%) of the seeds promoted 100% germination in this weedy species. That compares to 79% germination for the control.

Zaluzianskya villosa **(Thunb.) F. W. Schmidt**

BLUE DRUMSTICKS

This species occurs in Cape Floristic Region of South Africa. Treating its seeds with aerosol smoke significantly promoted germination (Brown et al. 2003; Brown and Botha 2004).

SELAGINELLACEAE

Selaginella uliginosa **(Labill.) Spring**

SWAMP SELAGINELLA

Soil samples from the Eden Burning Study Area, a dry sclerophyll forest in the Yalumba State Forest of New South Wales, Australia, were collected and air dried to test the effects of heat, smoke, and an interaction between the two cues on seeds from the seed bank. Samples exposed to heat treatment were incubated at 80°C for 60 minutes while those exposed to smoke were incubated in a room, where smoke was generated for 120 minutes. None of the treatments had any effect on the germination of the seeds of this species (Penman et al. 2008).

SOLANACEAE

Anthocercis littorea **Labill.**

YELLOW TAILFLOWER

Both smoke water and karrikinolide significantly improved germination in this species, which is commonly used in rehabilitation practices of disturbed areas of Western Australia (Commander et al. 2009a). An after-ripening treatment (warm, dry storage) had no effect on its seeds unless they were treated with karrikinolide (Commander et al. 2009b), after which germination was promoted. A 60-minute treatment with cold smoke had no effect on germination (Roche et al. 1997b).

Cyphanthera odgersii **ssp.** occidentalis **Haegi**

WESTERN WOOLLY CYPHANTHERA

WESTERN WOOLLY CYPHANTHERA

Western woolly cyphanthera, a rare and endangered native of the proteaceous heath, mixed eucalypt woodland, and mulga woodland communities of Western Australia, exhibited significantly increased germination when its seeds were exposed to a combination of pretreatments. The combination of pretreatments included nicking the seed coat to expose the endosperm, soaking the seeds in smoke water for 24 hours, and adding gibberellic acid (25 mg/L) to the germination medium (Cochrane et al. 2002).

Duboisia myoporoides **R. Br.**

CORKWOOD

Corkwood is a shrub or tree that occurs mainly on the east coast of Australia, especially on the fringes of rainforests. A 60 minute aerosol smoke treatment of soil samples containing seeds of this species, collected from forests in the Lamington National Park of Queensland, Australia, did not germinate in response to the treatment (Tang et al. 2003).

Nicotiana attenuata **Torr. ex Watson**

COYOTE TOBACCO

Coyote tobacco is a native component of sagebrush shrubland and pinyon-juniper woodlands in the western United States and northern Mexico. Preston and Baldwin (1999) reported that germination in this species increased to 100% when its seeds were soaked in a 1:300 dilution of smoke water. Less than 5% germination occurred in the control. Nitrogen oxides were unlikely to have played a role in the germination of coyote tobacco, which was probably due to unidentified cellulose combustion factors (Preston et al. 2004). Flematti et al. (2004) achieved 44% germination by treating the seeds with karrikinolide (100 ppb) in comparison to 33% in the control. Schwachtje and Baldwin (2004) suggested that the effects of smoke treatment increased gibberellic acid sensitivity, which correlated with decreased endogenous abscisic acid pools of dormant genotypes of *N. attenuata*. Krock et al. (2002) concluded that, in addition to smoke, abscisic acid and four terpenes that leach from litter of dominant vegetation helped with promoting germination in seeds of this post-fire annual species, which can lay dormant for over a century.

Nicotiana forsteri **Roem. & Schult.**

Nicotiana forsteri occurs mainly on the east coast of Australia, with naturalized populations in other parts of the world. A 60 minute aerosol smoke treatment of soil samples did not significantly promote germination in the species (Tang et al. 2003).

Nicotiana linearis **Phil.**

NICOTIANA

The seeds of this native species of the grasslands of northwestern Patagonian were treated with a combination of heat (80°C for 5 minutes) and aerosol smoke for 60 minutes with no significant effect to germination (Gonzalez and Ghermandi 2012).

Solanum aphyodendron **S. Knapp**

SOLANUM

A variety of different fire cues were tested on the seeds of this species, which were collected from a mixed forest located in a mountainous subtropical area of Mexico (Zuloaga-Aguilar et al. 2011). These included heat shock (100:15 sand: water (w:w) substrate at 120°C for 5 minutes), soaking the seeds in smoke water for 3 hours (prepared by burning 150 g of mixed forest litter and bubbling the resultant smoke into 1.5 L of distilled water, and adjusting the pH to 5 using sodium hydroxide) or ash (1.5 g of fine ash was added to the agar plates used to germinate the seeds). Combinations of these treatments were also tested for their effects on seed germination. The following treatments all significantly improved germination for this species: ash and heat shock applied on their own, as well as the combinations of heat shock and ash, and heat shock/ash/smoke water.

Solanum aviculare **G. Forst.**

KANGAROO APPLE

This member of the Solanaceae family occurs mainly on the east coast of Australia, with naturalized populations in other countries. Exposing soil samples containing seeds of this species to 60 minutes of aerosol smoke had no significant effect on germination (Tang et al. 2003).

Solanum centrale **J. M. Black**

DESERT RAISIN

Karrikinolide promoted seed germination in this native Australian species (Commander et al. 2008).

Solanum dioicum **W. Fitzg.**

Commander et al. (2008) reported that a smoke-isolated butenolide was effective in promoting seed germination in this native Australian species.

Solanum douglasii **Dunal**

WHITE NIGHTSHADE

Germination of white nightshade, a native of Mexico and southern states of the United States of America, ranging from California to Louisiana, was inhibited by 23% when powdered charred wood (0.5 g) was added to Petri dishes on which the seeds were tested (Keeley and Keeley 1987).

Solanum elaeagnifolium **Cav.**

SILVERLEAF NIGHTSHADE

Chou et al. (2012) tested the effects of smoke water, heat and combinations of them both on the seeds of this species. Smoke treatments comprised of soaking seeds for 20 hours in the commercially available Regen 2000® smoke water solution, at concentrations of 1:5, 1:10, or 1:100 (v/v). The seeds were then exposed heat shocked at 50 or 80°C for a period of 5 minutes. Neither smoke water nor heat treatments on their own significantly affected germination percentage or mean germination time, but there was an interaction between the two at the higher temperature and dilution of smoke water.

Solanum lycopersicum **L.**

TOMATO

Karrikinolide promoted the germination and seedling vigor of tomato seeds (Jain and van Staden 2007). In addition, it also ensured normal germination and seedling establishment at below (10°C) and above (40°C) optimum temperatures (Jain et al. 2006). Germination of the control seeds was optimal at 25°C. Taylor and van Staden (1998) tested smoke water on the growth of tomato roots. Primary root length and secondary root frequency increased significantly in response to increasing smoke water concentrations. The vigor index was significantly greater when karrikinolide-primed tomato seeds were germinated following different seed storage periods, and under different salt concentrations, osmotic potentials, and temperatures (Jain and van Staden 2007). Tomatoes treated with smoke water (1:500 v/v) achieved maximum height, number of leaves, and stem thickness 57 to 78 days after sowing. Following this treatment, there was a significant increase in the marketable fruits, with an average of 35% more fruits per plant (van Staden et al. 2008).

Solanum melongena **L.**

EGGPLANT

A smoke-derived karrikinolide, with and without priming agents, enhanced seedling size and promoted more rapid emergence in aged aubergine, *S. melongena* cv. "Kemer" seeds (Demir et al. 2009).

Solanum nigrum **L.**

BLACK NIGHTSHADE

This species has a broad native range throughout northern Africa, Europe,

Solanum nigrum

western and middle Asia, the Indian subcontinent, China, and Siberia. Black nightshade is a weed in many parts of the world. In an Australian study, Read et al. (2000) revealed that exposure to aerosol smoke for 60 minutes inhibited germination of its seeds.

Solanum orbiculatum **Poir**

WILD TOMATO

Flematti et al. (2007) and Commander et al. (2008) reported that karrikinolide and a number of its analogues were effective in promoting germination in this Western Australian species. Smoke water was also reported to be useful in promoting germination in this plant, which is often used in rehabilitation of disturbed areas (Commander at al. 2009a).

Solanum stelligerum **Sm.**

STAR NIGHTSHADE

Star nightshade is native to eastern Australia. A 60 minute aerosol smoke treatment of soil samples containing seeds of this species, collected across forest edges between subtropical rainforests and eucalypt forests in the Lamington National Park of Queensland, Australia, did not promote germination in the species (Tang et al. 2003).

Solanum viarum **Dunal**

TROPICAL SODA APPLE

Tropical soda apple is native to Brazil, Paraguay, Uruguay, and Argentina but has become a weed in South America, North America, India, and Africa. Several seed treatments, including mechanical and chemical scarification, cold stratification, gibberellic acid, nutrient solutions, smoke water, and karrikinolide were used to determine their effect on germination of this species (Kandari et al. 2011). We report only on the effects of smoke water and karrikinolide on the seed germination of this species. Smoke water was prepared by burning 5kg of *Themeda triandra* Forssk. (Poaceae) leaf material in a 20-L metal drum and pumping the resultant smoke through a glass column containing 500 mL of tap water for a period of 45 minutes. The seeds were then treated with three concentration of this smoke water solution (1:250, 1:500, and 1:1,000 (v/v) of smoke water: water). Concentrations of 10^{-7}, 10^{-8}, and 10^{-9} M of karrikinolide were prepared as described by van Staden et al. (2004). The percentage germination, germination rate, mean germination time, root length, shoot length, and seedling weight resulted in positive responses to smoke water or karrikinolide. Percentage germination and germination rate were significantly increased when seeds were treated with all smoke water or karrikinolide treatments. The mean germination time was significantly reduced as a result of all the treatments, with the exception of the 1:500 v/v smoke water treatment. The 1:500 v/v smoke water treatment resulted in a significantly greater root and shoot length compared to the control, whereas all other smoke water and karrikinolide treatments exhibited no significant effect. Seedling weight increased significantly in response to the 1:500 v/v smoke water and the 10^{-9} M karrikinolide treatments.

STACKHOUSIACEAE

Stackhousia huegelii **Endl.**

HEAVY-SCENTED STACKHOUSIA

Dixon et al. (1995) reported that 90 minutes of cold smoke treatment had no effect on seed germination of this Western Australian species.

Stackhousia monogyna **Labill.**

CREAMY STACKHOUSIA

This species occurs in the jarrah forests of Western Australia. Neither aerosol smoke nor smoke water treatments significantly promoted germination of its seeds (Norman et al. 2006).

Stackhousia pubescens **A. Rich.**

DOWNY STACKHOUSIA

Stackhousia pubescens is native to Western Australia, occurring in the proteaceous heath-scrub of the Southwest Botanical Province. Germination significantly increased in response to aerosol smoke applications (Dixon et al. 1995, Roche et al. 1997b). Aerosol smoke treatments of broadcast seeds in rehabilitated mine sites in the southwest of Western Australia had no effect on their germination, however (Roche et al. 1997a).

Tripterococcus brunonis **Endl.**

WINGED STACKHOUSIA

Tripterococcus brunonis has a broad distribution in Western Australia's southwest region (proteaceous heaths, banksia woodlands, mixed eucalypt woodlands, and jarrah forests). Seed germination was improved significantly when the soil seed bank was treated with smoke water (1 L/m²) but not when broadcast seeds in rehabilitated mine sites in the southwest of Western Australia were treated with aerosol smoke (Roche et al. 1997a). An ex situ study revealed maximum germination could be achieved if seeds were stored in soil for 12 months prior to their smoke application (Roche et al. 1997b). Interestingly, Norman et al. (2006) reported that neither aerosol smoke nor smoke water treatments significantly promoted germination in this species. Dixon et al. (1995) reported that 90 minutes of cold smoke treatment also had no effect on seed germination.

STERCULIACEAE

Fremontodendron californicum **(Torr.) Coville**

CALIFORNIA FLANNEL BUSH

California flannel bush is native to the chaparral, coastal scrub, and woodland communities of Arizona and California. Keeley (1987) was able to promote modest germination by using charate following a 5-minute heat shock treatment of 100°C or 120°C.

Fremontodendron decumbens **R. M. Lloyd**

PINE HILL FLANNEL BUSH

Pine hill flannel bush is rare and endangered (IUCN 2013) and restricted to California. Boyd and Serafini (1992) reported that powdered charred wood in the potting medium promoted seed germination in this species.

Hermannia alnifolia **L.**

Seed germination in this fynbos species of the Cape Floristic Region of South Africa was not significantly improved following 30 minutes of aerosol smoke treatment (Brown et al. 2003; Brown and Botha 2004).

Hermannia hyssopifolia **L.**

EIGHT-DAY HEALING BUSH

Seed germination in this fynbos species of the Cape Floristic Region of South Africa was not significantly improved following 30 minutes of aerosol smoke treatment (Brown et al. 2003; Brown and Botha 2004).

Hermannia rudis **N. E. Br.**

POPROSIES

Seed germination in this fynbos species of the Cape Floristic Region of South Africa was not significantly improved following 30 minutes of aerosol smoke treatment (Brown et al. 2003; Brown and Botha 2004).

Hermannia scabra **Cav.**

Seed germination in this fynbos species of the Cape Floristic Region of South Africa was not significantly improved following 30 minutes of aerosol smoke treatment (Brown et al. 2003; Brown and Botha 2004).

Hermannia **sp.**

Seed of an unidentified species of *Hermannia* of the Cape Floristic Region of South Africa did not significantly respond to 30 minutes of aerosol smoke treatment (Brown et al. 2003; Brown and Botha 2004).

Lasiopetalum floribundum **Benth.**

Lasiopetalum floribundum is native to proteaceous heathland, banksia woodland, and jarrah forest communities of the southwestern region of Western Australia. Germination in this species was promoted by ex situ aerosol smoke applications to the soil seed bank (Roche et al. 1997b). A 1 L/m² application of undiluted smoke water to the soil surface of a rehabilitated bauxite mine site had no significant effect on seed germination, however (Roche et al. 1997a).

Rulingia platycalyx **Benth.**

Rulingia platycalyx is commonly found growing in the proteaceous heath-scrub of the southwestern region of Western Australia. Germination in this species was significantly improved by aerosol smoke applications to the soil in which it had been sown (Dixon et al. 1995).

Thomasia glutinosa **Lindl.**

STICKY THOMASIA

Sticky thomasia is native to the proteaceous heathland, banksia woodland, and jarrah forest communities of Western Australia's southwest region. Germination was not significantly promoted by cool aerosol smoke treatments of freshly collected seeds (Roche et al. 1997b).

STILBACEAE

Stilbe vestita **P. J. Bergius**

Stilbe vestita is a component of the fynbos of Cape Floristic Region of South Africa. Brown et al. (2003) reported that germination significantly increased in response to aerosol smoke treatments.

STYLIDIACEAE

Levenhookia pusilla **R. Br.**

MIDGET STYLEWORT

Midget stylewort is native to the Southwest Botanical Province of Western Australia. Roche et al. (1997a) showed that germination of this species was significantly improved by an in situ application of aerosol smoke to the soil surface, but smoke water had no effect and nor did an aerosol smoke treatment of broadcast seeds in a rehabilitated mine site in the southwest corner of Western Australia.

Stylidium affine **Sond.**

QUEEN TRIGGERPLANT

The queen triggerplant occurs in the southwest regions of Western Australia, especially the proteaceous heathland, banksia woodland, and jarrah forest communities. Roche et al. (1997b) showed that optimal germination could be achieved if seed that had already been sown were treated with aerosol smoke for 60 minutes (see also Tieu et al. 2001b). Tieu et al. (1999) revealed that soaking the seeds in smoke water enhanced germination by breaking seed dormancy mechanisms. Karrikinolide significantly increased germination at concentrations ranging from 1 ppm to 1 ppt (Flematti et al. 2004). Germination was equivalent or higher than that obtained by treating the seeds with smoke water (1% or 10%). In contrast, Downes et al. (2010), who collected *S. affine* seeds from near Boddington, Western Australia, and exposed them to smoke water (1:10 [v/v]; pH = 4.05, "Seed Starter," Kings Park and Botanic Garden, Perth, Western Australia) and karrikinolide (KAR$_1$; 1 μM), revealed there was significantly increased germination following the smoke water treatment, but no effect when the seeds were tested with karrikinolide.

Stylidium amoenum **R. Br.**

LOVELY TRIGGERPLANT

This species commonly occurs in the proteaceous heathland, banksia woodland, and jarrah forest communities of Western Australia's southwestern region. A 60-minute treatment of its seeds significantly promoted germination (Roche et al. 1997b). Norman et al. (2007) reported that optimal germination of this species was achieved by immersing the seeds in smoke water (1%) and incubating them on an 18/10°C and light/dark 9/13 hour regime. Aerosol smoke treatments of broadcast seeds in rehabilitated mine sites in the southwest of Western Australia had no effect on their germination (Roche et al. 1997a).

Stylidium brunonianum **Benth.**

PINK FOUNTAIN TRIGGERPLANT

Stylidium brunonianum occurs in the southwest regions of Western Australia, especially the proteaceous heathland, banksia woodland, and jarrah forest communities. Roche et al. (1997b) showed that optimal germination could be achieved if seeds were treated with aerosol smoke for 60 minutes (see also Tieu et al. 2001b). Tieu et al. (1999) reported that soaking the seeds in smoke water enhanced germination by breaking its seed dormancy mechanisms. Smoke water sprays, with concentrations of 50 and 100 mL/m², did not, however, promote germination of seeds in an intact banksia woodland 20 km south of Perth, Western Australia (Lloyd et al. 2000).

Stylidium bulbiferum **Benth.**

CIRCUS TRIGGERPLANT

The circus triggerplant occurs in the southwestern regions of Western Australia, especially the proteaceous heathland, banksia woodland, and jarrah forest communities. Norman et al. (2006) reported that neither aerosol smoke nor smoke water treatments significantly promoted germination of its seed. Aerosol smoke treatments of broadcast seeds in rehabilitated mine sites in the southwest of Western Australia showed no effect on germination either (Roche et al. 1997a). Roche et al. (1997b) demonstrated, in contrast, that maximum germination could be achieved when the seeds were stored in soil for 12 months prior to treating them with aerosol smoke.

Stylidium calcaratum **R. Br.**

BOOK TRIGGERPLANT

Stylidium calcaratum is widely distributed throughout the Southwest Botanical Province of Western Australia. Roche et al. (1997a) reported that the in situ germination of this species was significantly promoted when the soil seed bank was treated with aerosol smoke, but smoke water had no effect, nor did an aerosol smoke treatment of broadcast seeds in rehabilitated mine sites in southwestern Western Australia. Norman et al. (2007) reported increased germination when seeds were treated with aerosol smoke (60 minutes) and incubated at 18/10°C and light/dark for 9/13 hours.

Stylidium crossocephalum **F. Muell.**

POSY TRIGGERPLANT

Posy triggerplant is native to the proteaceous heathlands, banksia woodlands, and jarrah forests of the Southwest Botanical Province of Western Australia. Tieu et al. (2001b) reported that dormant seeds required burial in soil for 450 days and smoke treatment for 60 minutes in order to achieve 88% germination.

Stylidium graminifolium **Sw. ex Willd.**

GRASS TRIGGERPLANT

Grass triggerplant is common in the dry sclerophyll forests of southeastern Australia. Germination in this species was 6.5% without treatment. Treating them with smoke water (10% dilution), chilling or heat did not promote any germination (Clarke et al. 2000).

Stylidium hispidum **Lindl.**

WHITE BUTTERFLY TRIGGERPLANT

White butterfly triggerplant occurs in the southwestern regions of Western Australia, especially the proteaceous heathland, banksia woodland, and jarrah forest communities. Roche et al. (1997b) reported that optimal germination could be achieved if seeds were treated with aerosol smoke for 60 minutes (see also Tieu et al. 2001b). Tieu et al. (1999) reported that soaking the seeds in smoke water also significantly promoted. Smoke water sprays, with concentrations of 50 and 100 mL/m², did not, however, promote germination of seeds in an intact banksia woodland 20 km south of Perth, Western Australia (Lloyd et al. 2000). A 1 L/m² application of undiluted smoke water to the soil surface of a rehabilitated bauxite mine in the southwest also exerted no significant effect on the germination of its seeds (Roche et al. 1997a).

Stylidium junceum **R. Br.**

REED TRIGGERPLANT

Refer to *S. affine* for similar distribution and smoke responses (Roche et al. 1997b). Smoke water (1%) also significantly improved germination of reed triggerplant under laboratory and field conditions (Norman et al. 2007). Smoke water had no effect (Roche et al. 1997a).

Stylidium repens **R. Br.**

MATTED TRIGGERPLANT

Matted triggerplant occurs primarily in southwest of Western Australia. Smoke water sprays, with concentrations of 50 and 100 mL/m^2, did not significantly promote germination of its seeds in an intact banksia woodland 20 km south of Perth, Western Australia (Lloyd et al. 2000).

Stylidium schoenoides **DC.**

COW KICKS

Germination of cow kicks, a native to southwestern Western Australia's proteaceous heathlands, banksia woodlands, and jarrah forests, increased in response to aerosol smoke treatments (Roche et al. 1997a, b, Norman et al. 2007). Smoke water (1%) also significantly improved germination of cow kicks under laboratory and field conditions (Norman et al. 2007).

Stylidium soboliferum **F. Muell.**

GRAMPIANS TRIGGERPLANT

Grampian's triggerplant occurs in the heath and eucalypt woodland communities of Victoria, Australia. Application of smoke water to the soil seed bank promoted germination in this species (Enright and Kintrup 2001).

THYMELAEACEAE

Gnidia pinifolia **L.**

Gnidia pinifolia is a native of South Africa. This species responded positively when its seeds were exposed to aerosol smoke for 30 minutes. Germination, however, remained low (5% final germination) (Brown and Botha 2004).

Passerina vulgaris **Thoday**

SAND GONNABOS

Sand gonnabos, a native of South Africa, germinated more readily when its seeds were exposed to aerosol smoke (Brown et al. 1995; see also Brown et al. 2003;

Brown and Botha 2004). Flematti et al. (2004) reported a mean germination percentage of 10% when they were treated with karrikinolide. The seeds did not germinate at all in the controls.

Pimelea ciliata **B. L. Rye**

WHITE BANJINE

The white banjine occurs in the southwest regions of Western Australia, especially the proteaceous heathland, banksia woodland, and jarrah forest communities. Roche et al. (1997b) showed that optimal germination could be achieved if sown seeds were treated with aerosol smoke for 60 minutes. Smoke water had no effect, nor did an aerosol smoke treatment of broadcast seeds of rehabilitated mines sites (Roche et al. 1997a). Norman et al. (2006) reported that neither aerosol smoke nor smoke water treatments significantly promoted germination of its seeds.

Pimelea imbricata **R. Br.**

Pimelea imbricata is widely distributed throughout the Southwest Botanical Province of Western Australia. Seed germination of this species was higher when sown seeds were exposed to aerosol smoke for a period of 60 minutes (Roche et al. 1997b).

Pimelea leucantha **Diels**

Pimelea leucantha is native to the southwestern Western Australian proteaceous heath-scrub communities. Roche et al. (1997b) achieved 18% germination by treating seeds that had been stored in soil for 12 months with aerosol smoke for 60 minutes. Dixon et al. (1995) reported that 90 minutes of cold smoke treatment had no effect on seed germination.

Pimelea **sp.**

A 60 minute aerosol smoke treatment of soil samples containing seeds of an unidentified species of *Pimelea*, collected across forest edges between subtropical rainforests and eucalypt forests in the Lamington National Park of Queensland, Australia, did not germinate (Tang et al. 2003).

Pimelea spectabilis **Lindl.**

BUNJONG

Refer to *Lechenaultia floribunda* (Epacridaceae) for similar distribution and smoke responses (Dixon et al. 1995). Aerosol smoke treatments of broadcast seeds in rehabilitated mine sites in the southwest of Western Australia had no effect on their germination (Roche et al. 1997a).

Pimelea spicata **R. Br.**

RICE FLOWER

This endangered (IUCN 2013) endemic shrub is restricted to 30 populations in the Cumberland Plain Woodland community of New South Wales, Australia. Willis et al. (2003) showed that a 30% increase in germination occurred when its seeds were exposed for 16 minutes to aerosol smoke.

Pimelea suaveolens **Meisn.**

SCENTED BANJINE

Scented banjine is native to the Western Australian proteaceous heathland, banksia woodland, jarrah forest, and mixed eucalypt woodland habitats. Norman et al. (2006) reported that neither aerosol smoke nor smoke water treatments significantly promoted germination of its seed. Roche et al. (1997a) discovered, in contrast, that seed germination was significantly enhanced when (1) concentrated smoke water was applied to the soil seed bank at a rate of 1 L/m^2, and (2) when seeds were treated with aerosol smoke for 60 minutes prior to sowing. Ex situ germination trials were also conducted and revealed that treatment with aerosol smoke for 60 minutes, after the seeds were sown, also improved germination (Roche et al. 1997b).

Pimelea sylvestris **R. Br.**

Pimelea sylvestris occurs in the proteaceous heathland, banksia woodland, and jarrah forest regions of Western Australia's southwestern region. Ex situ germination was significantly improved when the soil seed bank was exposed to aerosol smoke for 90 minutes (Dixon et al. 1995). Aerosol smoke treatments of broadcast seeds in rehabilitated mine sites in the southwest of Western Australia had no effect on their germination (Roche et al. 1997a).

Struthiola myrsinites **Lam.**

Struthiola myrsinites is a native of the South African fynbos. Germination increased 330% in response to aerosol smoke treatments of 30 minutes duration (Brown and Botha 2004).

TREMANDRACEAE

Tetratheca hirsuta **Lindl.**

BLACK-EYED SUSAN

The germination of *T. hirsuta*, a native of Western Australia's proteaceous heathland, banksia woodland, and jarrah forest communities, increased from a mean of 2.67 germinants to a mean of 6.93 germinants when the soil seed bank was treated with smoke water (Roche *et al.* 1997a). In addition, germination increased from a mean of 5.5 germinants to 11.5 germinants when the seeds were pretreated with 60 minutes of aerosol smoke. This study was conducted on a bauxite mine site in the process of being rehabilitated with native vegetation. Ex situ seed treatments with aerosol smoke applications to the soil surface also exhibited an increase in germination (Dixon et al. 1995; Roche et al. 1997b). Norman et al. (2006) reported, in contrast, that neither aerosol smoke nor smoke water treatments significantly promoted germination in this species.

Tetratheca pilosa **Labill.**

HAIRY PINK BELLS

Soil samples from the Eden Burning Study Area, a dry sclerophyll forest in the Yalumba State Forest of New South Wales, Australia, were collected and air dried to test the effects of heat, smoke, and an interaction between the two cues on seeds from the seed bank. Samples exposed to heat treatment were incubated at 80°C for 60 minutes while those exposed to smoke were incubated in a room, where smoke was generated for 120 minutes. Both heat and smoke treatments on their own significantly improved germination of the seeds of this species while an interaction between the two cues had no effect (Penman et al. 2008).

ULMACEAE

Trema tomentosa **(Roxb.) Hara**

POISON PEACH

Poison peach is native to several countries, including parts of Africa and Asia, and along the east coast of Australia. A 60 minute aerosol smoke treatment of soil samples containing seeds of this species, collected across forest edges between subtropical rainforests and eucalypt forests in the Lamington National Park of Queensland, Australia, had no effect on germination (Tang et al. 2003).

URTICACEAE

Dendrocnide excelsa **(Wedd.) Chew**

GIANT STINGING TREE

The giant stinging tree is native to eastern Australia. A 60 minute aerosol smoke treatment of seeds collected in Queensland, Australia, did not significantly promote germination (Tang et al. 2003).

Urtica incisa **Poir**

SCRUB NETTLE

The scrub nettle is native to the southeastern Australia. A 60 minute aerosol smoke treatment of soil samples containing the seeds of this species, collected from the Lamington National Park of Queensland, Australia, did not significantly affect germination (Tang et al. 2003).

VERBENACEAE

Lantana camara **L.**

LANTANA

Lantana is native to the American tropics and has naturalized in many parts of the world through its distribution as an ornamental plant. A 60 minute aerosol smoke treatment of soil samples containing seeds of this species, collected from the Lamington National Park of Queensland, Australia, did not germinate in response to the treatment (Tang et al. 2003).

Tectona grandis **L. f.**

TEAK

In this study, the seeds of teak were collected from a dry deciduous forest in India and later exposed to aerosol smoke until they turned completely brown. The seeds were arranged onto filter papers that were hung in fumigation tents, in which litter collected from locally grown dry deciduous trees was burned to generate the smoke. Germination percentages and the germination velocity index of seeds were measured (Singh and Raizada 2010). The results revealed that germination percentage (37%) and germination velocity index (0.85) were both significantly increased following the smoke treatment. This was compared to 25% and 0.53, respectively, for the control group. Teak is an economically important tree that has a number of uses, including the manufacture of musical instruments.

VIOLACEAE

Hybanthus calycinus **F. Muell.**

WILD VIOLET

Germination of this native of southwestern Western Australian proteaceous heathland, banksia woodland, and jarrah forest communities, was significantly increased if its seeds had been stored in soil for 12 months, after which aerosol smoke was applied to it for 60 minutes (Roche et al. 1997b). Aerosol smoke treatments of broadcast seeds in rehabilitated mine sites in the southwest of Western Australia had no effect on their germination (Roche et al. 1997a).

Hybanthus enneaspermus **(L.) F. Muell.**

SPADE FLOWER

Spade flower occurs in several parts of Australia. A 60 minute aerosol smoke treatment of soil samples containing seeds of this species, collected from the Lamington National Park of Queensland, Australia, did not germinate (Tang et al. 2003).

Hybanthus floribundus **(Lindl.) F. Muell.**

SHRUB VIOLET

Hybanthus floribundus is a native of southwestern Western Australia's proteaceous heathland, banksia woodland, and jarrah forest regions. Ex situ germination significantly improved, especially when the soil seed bank was treated with aerosol smoke for 90 minutes (Dixon et al. 1995). Roche et al. (1997b) and Norman et al. (2006) reported that neither aerosol smoke nor smoke water treatments significantly promoted germination in this species.

Viola betonicifolia **Sm.**

ARROWHEAD VIOLET

The arrowhead violet occurs in India, Pakistan and parts of Australia. A 60 minute aerosol smoke treatment of soil samples containing seeds of this species, collected from the Lamington National Park of Queensland, Australia, did not germinate (Tang et al. 2003).

Viola hederacea **Labill.**

NATIVE VIOLET

Like *Viola betonicifolia* above, the native violet is also native to Australia. There was no significant effect on seed germination following a 60-minute aerosol smoke treatment of soil samples containing the seeds of this species. This trial was performed in the Lamington National Park of Queensland, Australia (Tang et al. 2003). In another study, soil samples from the Eden Burning Study Area, a dry sclerophyll forest in the Yalumba State Forest of New South Wales, Australia, were collected and air dried to test the effects of heat, smoke, and an interaction between the two cues on seeds from the seed bank (Penman et al. 2008). Samples exposed to heat treatment were incubated at 80°C for 60 minutes while those exposed to smoke were incubated in a room, where smoke was generated for 120 minutes. None of the treatments had any effect on the germination of the seeds of this species.

VITACEAE

Ampelopsis glandulosa **(Wall.) Momiy. var.** heterophylla **(Thunb.) Momiy.**

FLIKBLADIGT GLASPÄRLEVIN (SWEDISH)

The effects of aerosol smoke, heat, darkness, cold stratification, and combinations of smoke with each of the three other treatments on seed germination were examined in this study (Tsuyuzaki and Miyoshi 2008). Smoke was produced by burning Timothy hay (*Phleum pratense*), which was pumped through a 3.5 m cooling tube into a smoke chamber for approximately 5 minutes. The seeds were exposed to the smoke for 60 minutes. Those seeds exposed also to heat were incubated at 75°C for 25 minutes. The cold stratification process took 1 month, during which the seeds remained in an incubator set to 4°C. Where the dark treatment was concerned, the seeds were maintained in total darkness for the entire germination period. Treatment under dark conditions resulted in 23% germination. There was 92% germination following cold stratification and 86% when after the seeds had received the combined smoke and cold stratification treatment. Germination in the control group was 48%. This species occurs in the temperate regions of Asia.

Tetrastigma nitens **(F. Muell.) Planch.**

SHINING GRAPE

The shining grape is endemic to Australia. A 60 minute aerosol smoke treatment of seeds collected from the Lamington National Park of Queensland, Australia, had no significant effect on germination (Tang et al. 2003).

Vitis coignetiae **Pulliat ex. Planch.**

CRIMSON GLORY VINE

The effects of aerosol smoke, heat, darkness, cold stratification, and combinations of smoke with each of the three other treatments on seed germination were examined in this study (Tsuyuzaki and Miyoshi 2008; see *Ampelopsis glandulosa* var. *heterophylla* above for details about the tests performed). Germination in the controls was negligible (0%–0.5%), but significantly improved to 11% germination when the seeds were cold stratified. These seeds display physiological dormancy. This species occurs in the temperate climes of Asia.

GLOSSARY OF TERMS AND ABBREVIATIONS

angiosperm: A seed bearing plant, in which the ovule is encased within an ovary, e.g., magnolias and other flowering plants.

biodiversity: The totality of genes, species, and ecosystems of a geographical region.

biomes: Biomes are essentially ecosystems whose boundaries are defined by assemblages of the distinct organisms and communities within them.

biosynthesis: The buildup of chemical compounds using simpler "building blocks" (e.g., amino acids, simple sugars), which is usually catalyzed by an enzyme.

butenolides: A class of compounds that promotes germination in the seed of certain species of plant. See also karrikinolide.

callogenesis: The formation of an undifferentiated mass of plant cells, called calli.

chaparral: A biome in the western United States, characterized by hot dry summers and cool moist winters. It is usually dominated by a dense growth of mostly small-leaved evergreen shrubs rich in highly flammable resins, which aid the frequent wildfires.

charate: The products of powdered, charred wood.

congeneric: Belonging to the same genus.

cotyledon: The leaves of an embryonic plant. These usually emerge shortly after the plant has sprouted. Monocotyledons produce one leaf, while dicotyledons produce two.

DNA: Deoxyribonucleic acid. A nucleic acid with genetic instructions that is used for the creation and functioning of all known organisms and some viruses.

dormancy: A state of quiet or temporary inaction. Where seed germination is concerned, it refers to a seed that does not have the capacity to germinate in a specified period of time under any combination of normal physical environmental factors, which are otherwise favorable for its germination (Baskin and Baskin 2004).

ecoregion: Also known as an ecological region or bioregion, ecoregions are large expanses of land, water, or both combined, that are larger than ecosystems and smaller than ecozones, and which possess distinct assemblages of organisms and communities.

ectoparasite: Any parasite that lives on the exterior of its host.

endemic: Native to a limited region or area.

expansins: A group of nonenzymatic proteins involved in a variety of important roles in plants, especially where cell loosening occurs, e.g., cell growth, seed germination, fruit softening, and many others. They are usually located in plant cell walls.

fynbos: The natural shrubland vegetation that occurs in the Western Cape floristic region of South Africa. It is characterized by winter rainfall in coastal and mountainous areas with a Mediterranean climate. Fire is a necessary component in the life cycles of almost all fynbos plant species.

GA3: Gibberellic acid.

genetic erosion: Loss of genetic diversity in a population or sample from random or selective processes.

germination: The process during which seeds begin to grow.

gymnosperms: Plants that do not possess ovules encased within an ovary, e.g., pine trees.

heathland: An ecosystem characterized by low evergreen shrubs of the family Ericaceae.

holarctic: Refers to ecozones found throughout the northern continents of the world.

HSPs: Heat shock proteins. These proteins are produced in plants in response to high temperatures.

in situ: In a natural habitat.

invasive plant: A plant that grows outside of its native range and tends to outcompete native flora.

in vitro: A procedure performed outside of a living organism, usually in a test tube, Petri dish, or other container.

IUCN: International Union for Conservation of Nature and Natural Resources.

karrikinolide: The active substance in smoke that promotes germination. It belongs to the class of substances known as the butenildes. Its chemical name is 3-methyl-$2H$-Furo[2,3-c]pyran-2-one. The word, karrikinolide, was derived from the Australian Aboriginal Nyungar word for smoke, *karrick*.

karrikins: Analogous butenolide substances.

leachate: A solution that has leached from a solid, e.g., leaves and other plant parts.

matorral: A Spanish word meaning shrubland or thicket. In Chile, the matorral is a terrestrial ecoregion comprised of Mediterranean forests, scrub, and woodlands.

MD: Morphological dormancy.

MDEs: microspore-derived embryos.

MPD: Morphophysiological dormancy.

mg: Miligram. One-thousandth of a gram.

moiety: In reference to a molecule. A "moiety" is a part or section of the whole.

monocarpic: Plants that flower only once.

natural selection: Genes that produce characteristics more favorable to a particular environment will be more abundant in the next generation.

nGA$_3$: Gibberellic acid nitrite.

NO: Nitrogen oxides.

O$_2$: Oxygen gas.

obligate seeder: Recovery after fire occurs only through seed recovery as opposed to vegetative sprouting.

paleoclimate: The climate of a given period of time in the geologic past.

PD: Physiological dormancy.

photodormancy: A type of dormancy displayed by seeds that require a high amount of unfiltered solar radiation in order to germinate.

phylogeny: The evolutionary relationships between and among organisms.

ppm: Parts per million.

ppt: Parts per thousand.

priming: Pretreatment of seeds by various methods in order to improve seed germination.

proteaceous: In reference to shrubs, native to Australia or South Africa, within the family Proteaceae.

PSV: Protein storage vacuoles.

PY: Physical dormancy.

scarification: The process of disrupting a seed coat to encourage germination.

sclerophyllous woodlands: Woodlands dominated by plants whose leaves are hard, thick, leathery, and often small, allowing for the plants to survive hot, dry Mediterranean-like climates.

seed bank: A store of viable seed dormant in the soil or underwater.

solanaceous plants: Plants belonging to the family Solanaceae.

ssp: Subspecies.

syn: Synonym. In botany, a synonym is an illegitimate name that has been rejected and changed for a more appropriate one. It is common for synonyms, when they exist, to also be included in literature describing plant species.

taxon: A taxonomic category into which related organisms are classified. The plural is taxa.

var: Variety.

vernacular: Common name.

vernalization: Prolonged exposure to cold treatment.

weed: A plant that grows outside of its native range and tends to outcompete native flora.

REFERENCES

Abdollahi, M. R., P. Ghazanfari, P. Corral-Martínez, A. Moieni, and J. M. Seguí-Simarro. 2012. Enhancing secondary embryogenesis in *Brassica napus* by selecting hypocotyl-derived embryos and using plant-derived smoke extract in culture medium. *Plant Cell, Tissue and Organ Culture* 110: 307–315.

Abella, S. R. 2006. Effects of smoke and fire-related cues on *Penstemon barbatus* seeds. *American Midland Naturalist* 155: 404–410.

Abella, S. R., and Springer, J. D. 2009. Planting trials in northern Arizona ponderosa pine forests. *Ecological Restoration* 27: 290–299.

Adkins, S. W., and N. C. B. Peters. 2001. Smoke derived from burnt vegetation stimulates germination of arable weeds. *Seed Science Research* 11(3): 213–222.

Adkins, S. W., and J. D. Ross. 1981. Studies in wild oat seed dormancy. I. The role of ethylene in dormancy breakage and germination of wild oat seeds (*Avena fatua* L.). *Plant Physiology* 67: 358–362.

Allan, S. M., S. W. Adkins, C. A. Preston, and S. M. Bellairs. 2004. Improved germination of the Australian natives: *Hibbertia commutata*, *Hibbertia amplexicaulis* (Dilleniaceae), *Chameascilla corymbosa* (Liliaceae) and *Leucopogon nutans* (Epacridaceae). *Australian Journal of Botany* 52(3): 345–351.

Anderson, T. M., M. Schutz, and A. C. Risch. 2012. Seed germination cues and the importance of the soil seed bank across an environmental gradient in the Serengeti. *Oikos* 121: 306–312.

Baker, K. S., K. J. Steadman, J. A. Plummer, and K. W. Dixon. 2005. Seed dormancy and germination responses of nine Australian fire ephemerals. *Plant and Soil* 277: 345–358.

Baldos, O. C., J. DeFrank, and G. S. Sakamoto. 2011. Improving germination of piligrass (*Heteropogon contortus*) seeds using liquid smoke flavouring. 2011 *American Society of Horticultural Science Annual Conference-Poster Session Abstracts*. September 25–28, 2011. Waikoloa, Hawaii.

Baldwin I. T., L. Staszak-Kozinski, and R. Davidson. 1994. Up in smoke. I. Smoke-derived germination cues for post-fire annual, *Nicotiana attenuata* Torr ex Watson. *Journal of Chemical Ecology* 20: 2345–2371.

Bar Nun, N., and A. M. Mayer. 2005. Smoke chemicals and coumarin promote the germination of the parasitic weed *Orobanche aegyptia*ca. *Israel Journal of Plant Sciences* 53(2): 97–101.

Baskin, C. C., and J. M. Baskin. 1998. *Seeds: ecology, biogeography, and evolution of dormancy and germination*. New York: Academic Press.

Baskin, J. M., and C. C. Baskin. 2004. A classification system for seed dormancy. *Seed Science Research* 14: 1–16.

Bassi, P. K., E. B. Tregunna, and A. N. Purohit. 1975. Carbon dioxide requirements for phytochrome action in photoperiodism and seed germination. *Plant Physiology* 56: 335–336.

Baxter, B. J. M., J. van Staden, J. E. Granger, and N. A. C. Brown. 1994. Plant-derived smoke and smoke extracts stimulate seed germination of the fire-climax grass *Themeda triandra*. *Environmental and Experimental Botany* 34: 217–223.

Bean, A. R. 2008. A synopsis of *Ptilotus* (Amaranthaceae) in eastern Australia. *Telopia* 12(2): 227–250.

Bell, T. L., J. S. Pate, and K. W. Dixon. 1996. Relationships between fire response, morphology, root anatomy and starch distribution in south-west Australian Epacridaceae. *Annals of Botany* 77(4): 357–364.

Bell, D. T., S. Vlahos, and L. E. Watson. 1987. Stimulation of seed germination of understorey species of the northern jarrah forest of Western Australia. *Australian Journal of Botany* 35: 593–599.

Berrie, A. M. M., M. R. Hendrie, W. Parker, and B. A. Knights. 1967. Induction of light sensitive dormancy in seed of *Lactuca sativa* L. (lettuce) by patulin. *Plant Physiology* 42: 889–890.

Bethke, P. C., I. G. L. Libourel, N. Aoyama, Y. Chung, D. W. Still, and R. L. Jones. 2007. The Arabidopsis Aleurone Layer Responds to Nitric Oxide, Gibberellin, and Abscisic Acid and Is Sufficient and Necessary for Seed Dormancy. *Plant Physiology* 143: 1173–1188.

Bond, W. J., and B. W. van Wilgen. 1996. Fire and Plants. London, UK: Chapman and Hall.

Boston, R. S., P. V. Viitanen, and E. Vierling. 1996. Molecular chaperones and protein folding in plants. *Plant Molecular Biology* 32: 191–222.

Botanic Gardens Trust. 2013. PlantNET—The Plant Information Network System of Botanic Gardens Trust. Retrieved 2013 from http://plantnet.rbgsyd.nsw.gov.au.

Boyd, R. S., and L. L. Serafini. 1992. Reproductive attrition in the rare chaparral shrub *Fremontodendron decumbens* Lloyd (Sterculiaceae). *American Journal of Botany* 79: 1262–1272.

Brown, N. A. C. 1993a. Promotion of germination of fynbos seeds by plant-derived smoke. *New Phytologist* 123: 575–583.

———. 1993b. Seed germination in the fynbos fire ephemeral, *Syncarpha vestita* (L.) B. Nord is promoted by smoke, aqueous extracts of smoke and charred wood derived from burning the ericoid-leaved shrub, *Passerina vulgaris* Thoday. *International Journal of Wildland Fire* 3: 203–206.

Brown, N. A. C., and P. A. Botha. 2004. Smoke seed germination studies and a guide to seed propagation of plants from the major families of the Cape Floristic Region, South Africa. *South African Journal of Botany* 70(4): 559–581.

Brown, N. A. C., P. A. Botha, and D. S. Prosch. 1995. Where there's smoke. *Journal of the Royal Horticultural Society* 120: 402–405.

Brown, N. A. C., H. Jamieson, and P. A. Botha. 1994. Stimulation of seed germination in South African species of Restionaceae by plant-derived smoke. *Plant Growth Regulation* 15(1): 93–100.

Brown, N. A. C., G. Kotze, and P. A. Botha. 1993. The promotion of seed germination of Cape Erica species by plant-derived smoke. *Seed Science and Technology* 21: 179–185.

Brown, N. A. C., D. S. Prosch, and P. A. Botha. 1998. Plant derived smoke: An effective pre-treatment for seeds of *Syncarpha* and *Rhodocoma* and potentially for many other Fynbos species. *South African Journal of Botany* 64(1): 90–92.

Brown, N. A. C., J. van Staden, M. I. Daws, and T. Johnson. 2003. Patterns in the seed germination response to smoke in plants from the Cape Floristic Region, South Africa. *South African Journal of Botany* 69(4): 514–525.

Brummitt, R. K., and C. E. Powell.eds. 1992. *Authors of plant names*. Richmond, Surrey, UK: Royal Botanic Gardens, Kew.

Burmeister, H. R., J. J. Ellis, and S. G. Yates. 1971. Correlation of biological to chromatographic data for two mycotoxins elaborated by *Fusarium*. *Applied Microbiology* 21(4): 673–675.

Butler, L. G. 1995. Chemical communication between the parasitic weed Striga and its crop host. A new dimension in allelochemistry. In K. M. Inderjit, M. Dashini, and F. A. Enhelling. 1995. *Allelopathy, organisms, procedures and applications*. Washington, DC: American Chemical Society.

Cadman, C. S. C., P. E. Toorop, H. W. M. Hilhorst, and W. E. Finch-Savage. 2006. Gene expression profiles of Arabidopsis Cvi seeds during dormancy cycling indicate a common underlying dormancy control mechanism. *The Plant Journal* 47(1): 164.

Campbell, S. D. 1995. Plant mechanisms that influence the balance between *Heteropogon contortus* and *Aristida ramosa* in spring burnt pastures. PhD thesis, University of Queensland, Brisbane, Australia.

Çatav, S. S., I. Bekar, B. S. Ateş, G. Ergan, F. Oymak, E. D. Ulker, and C. Tavşanoğlu. 2012. Germination response of five eastern Mediterranean woody species to smoke solutions derived from various plants. *Turkish Journal of Botany* 36(2012): 480–487.

Chen, F., and K. J. Bradford. 2000. Expression of an expansin is associated with endosperm weakening during tomato seed germination. *Plant Physiology* 124: 1265–1274.

Chen, C., J. Zou, S. Zhang, D. Zaitlin, and L. Zhu. 2009. Strigolactones are a new-defined class of plant hormones which inhibit shoot branching and mediate the interaction of plant-AM fungi and plant-parasitic weeds. *Science in China Series C: Life Sciences* 52(8): 693–700.

Chou, Y., R. D. Cox, and D. B. Wester. 2012. Smoke water and heat shock influence germination of shortgrass prairie species. *Rangeland Ecology and Management* 65(3):260–267.

Chumpookam, J., L. Huey-Ling, and S. Ching-Chang. 2012. Effect of smoke-water on seed germination and seedling growth of papaya (*Carica papaya* cv. Tainung No. 2). *Hortscience* 47(6): 741–744.

Clarke, P. J., E. A. Davison, and L. Fulloon. 2000. Germination and dormancy of grassy woodlands and forest species: effects of smoke, heat, darkness and cold. *Australian Journal of Botany* 48: 687–700.

Clarke, S., and K. French. 2005. Germination response to heat and smoke of 22 Poaceae species from grassy woodlands. *Australian Journal of Botany* 53(5): 445–454.

Coates, T. D. 2003. The effect of concentrated smoke products on the restoration of highly disturbed mineral sands in southeast Victoria. *Ecological Management and Restoration* 4(2): 133–139.

Cochrane, A., A. Kelly, K. Brown, and S. Cunneen. 2002. Relationships between seed germination requirements and ecophysiological characteristics aid the recovery of threatened native plant species in Western Australia. *Ecological Management and Restoration* 3(1): 47–60.

Commander, L. E., D. J. Merritt, D. P. Rokich, and K. W. Dixon. 2009a. Seed biology of Australian arid zone species: Germination of 18 species used for rehabilitation. *Journal of Arid Environments* 73: 617–625.

———. 2009b. The role of after-ripening in promoting germination of arid zone seeds: a study on six Australian species. *Botanical Journal of the Linnean Society* 161(4): 411–421.

Commander, L. E., D. J. Merritt, D. P. Rokich, G. R. Flematti, and K. Dixon. 2008. Seed germination of *Solanum* spp. (Solanaceae) for use in rehabilitation and commercial industries. *Australian Journal of Botany* 56(4): 333–341.

Conservation International, February, 2013. http://www.conservation.org/Pages/default.aspx#.

Crosti, R., P. G. Ladd, K.W. Dixon, and B. Piotto. 2006. Post-fire germination: The effect of smoke on seeds of selected species from the central Mediterranean basin. *Forest Ecology and Management* 221(1–3): 306–312.

Davidson, P. J., and S. W. Adkins. 1997. Germination of Triodia grass seeds by plant-derived smoke. In E. D. Mull, M. J. Page, and B. Alchin eds. Proceedings of the Australian Rangelenad Society 10th biennial Conference, December 1–10, 1997, Gatton College, Gatton, Australia.

Daws, M. I., J. Davies, H. W. Pritchard, N. A. C. Brown, and J. Van Staden. 2007. Butenolide from plant-derived smoke enhances germination and seedling growth of arable weed species. *Plant Growth Regulation* 51(1): 73–82.

Daws, M. I., H. W. Pritchard, and J. Van Staden. 2008. Butenolide from plant-derived smoke functions as a strigolactone analogue: Evidence from parasitic weed seed germination. *South African Journal of Botany* 74: 116–120.

Dayamba, S. D., L. Sawadogo, M. Tigabu, P. Savadogo, D. Zida, D. Tiveau, and P. C. Oden. 2010. Effects of aqueous smoke solutions and heat on seed germination of herbaceous species of the Sudanian savanna-woodland in Burkina Faso. *Flora* 205: 319–325.

Dayamba, S. D., M. Tigabu, L. Sawadogo, and P. C. Oden. 2008. Seed germination of herbaceous and woody species of the Sudanian savanna-woodland in response to heat shock and smoke. *Forest Ecology and Management* 256(3): 462–470.

DeRocher, A. M., and E. Vierling. 1994. Developmental control of small heat shock protein expression during pea seed maturation. *The Plant Journal* 5(1): 93–102.

Demir, I., M. E. Light, J. van Staden, B. B. Kenanoglu, and T. Celikkol, T. 2009. Improving seedling growth of unaged and aged aubergine seeds with smoke-derived butenolide. *Seed Science and Technology* 37(1): 255–260.

de Lange J. H., and C. Boucher. 1990. Autoecological studies on *Audouinia capitata* (Bruniaceae). I. Plant-derived smoke as a seed germination cue. *South African Journal of Botany* 56: 700–703.

Dhindwal, A. S., B. P. S. Lather, and J. Singh. 1991. Efficacy of seed treatment on germination, seedling emergence and vigour of cotton (*Gossypium hirsutum*) genotypes. *Seed Research* 19: 59–61.

Dixon K. W., S. Roche, and J. S. Pate. 1995. The promotive effect of smoke derived from burnt native vegetation on seed germination of Western Australian plants. *Oecologia* 101: 185–192.

Djietror, J. C., M. Ohara, and C. Appiah. 2011. Predicting the establishment and spread of siam weed in Australia. A test of abiotic cues on seed dormancy and germination. *Research Journal of Forestry* 2011: 1–13.

Doherty, L. C., and M. A. Cohn. 2000. Seed dormancy in red rice (*Oryza sativa*). XI. Commercial liquid smoke elicits germination. *Seed Science Research* 10: 415–421.

Downes, K. S., B. B. Lamont, M. E. Light, and J. van Staden. 2010. The fire ephemeral *Tersonia cyathiflora* (Gyrostemonaceae) germinates in response to smoke but not the butenolide 3-methyl-2H-furo[2,3-c]pyran-2-one. *Annals of Botany* 106: 381–384.

Downes, K. S., M. E. Light, M. Posta, L. Kahout, and J. van Staden. 2013. Comparison of germination responses of *Anigozanthos falvidus* (Haemodoraceae), *Gyrostemon racemiger* and *Gyrostemon ramulosus* (Gyrostemonaceae) to smoke-water and the smoke-derived compounds karrikinolide (KAR₁) and glyceronitrile. *Annals of Botany* 111(3): 489–497.

Drewes, F. E., M.T. Smith, and J. van Staden. 1995. The effect of plant-derived smoke extract on the germination of light-sensitive lettuce seed. *Plant Growth Regulation* 16: 205–209.

EcoPort. 2013. Ecoport. Retrieved 2013 from http://www.ecoport.org.

Enright, N. J., and A. Kintrup. 2001. Effects of smoke, heat and charred wood on the germination of dormant soil-stored seeds from a *Eucalyptus baxteri* heathy-woodland in Victoria, SE Australia. *Austral Ecology* 26: 132–141.

Finch-Savage, W. E., and G. Leubner-Metzger. 2006. Tansley Review: Seed dormancy and the control of germination. *New Phytologist* 171: 501–523.

Flematti, G. R., E. L. Ghisalberti, K. W. Dixon, and R. D. Trengove. 2004. A compound from smoke that promotes seed germination. *Science* 305: 977.

———. 2005. Synthesis of the seed germination stimulant 3-methyl-2*H*-furo[2,3-*c*] pyran-2-one. *Tetrahedron Letters*. 46(34): 5719–5721.

———. 2009. Identification of alkyl substituted 2*H*-Furo[2,3-*c*]pyran-2-ones as germination stimulants present in smoke. *Journal of Agricultural and Food Chemistry* 57(20): 9475–9480.

Flematti, G. R., E. D. Goddard-Borger, D. J. Merritt, E. L. Ghisalberti, K. W. Dixon, and R. D. Trengove. 2007. Preparation of 2*H*-Furo[2,3-*c*]pyran-2-one derivatives and evaluation of their germination-promoting activity. *Journal of Agricultural and Food Chemistry* 55: 2189–2194.

Flemmatti, G. R., D. J. Merritt, M. J. Piggott, R. D. Trengove, S. M. Smith, K. W. Dixon, and E. L. Ghisalberti. 2011a. Burning vegetation produces cyanohydrins that liberate cyanide and stimulate seed germination. *Nature Communications*. 2: 360.

Flemmatti, G. R., A. Scaffidi, K. W. Dixon, S. M. Smith, and E. L. Ghisalberti. 2011b. Production of the seed germination stimulant karrikinolide from combustion of simple carbohydrates. *Journal of Agricultural and Food Chemistry* 59: 1195–1198.

Fotheringham, C. J., E. Siu, and J. E. Keeley. 1995. Smoke stimulated germination in chaparral and coastal sage species of California. *Bulletin of the Ecological Society of America* 76(3): 327–328.

Franzese, J., and L. Ghermandi. 2011. Seed longevity and fire: germination responses of an exotic perennial herb in NW Patagonian grasslands (Argentina). *Plant Biology* 13: 865–871.

Frazer, J. G. 1922. *The golden bough: a study in magic and religion.* New York: Macmillan.

Gerstner, J. 1939. A preliminary checklist of Zulu names of plants with short notes. *Bantu Studies* 13(1): 49–64, 131–149.

Ghazanfari, P., M. R. Abdollahi, A. Moieni, and S. S. Moosavi. 2012. Effect of plant-derived smoke extract on in vitro plantlet regeneration from rapeseed (*Brassica napus* L. cv. Topas) microspore-derived embryos. *International Journal of Plant Production* 6(3): 309–323.

Ghebrehiwot, H. M., M. G. Kulkarni, K. P. Kirkman, and J. Van Staden. 2008. Smoke-water and a smoke-isolated butenolide improve germination and seedling vigour of

Eragrostis tef (Zucc.) Trotter under high temperature and low osmotic potential. *Journal of Agronomy and Crop Science* 94(4): 270–277.

————. 2009. Smoke solutions and temperature influence the germination and seedling growth of South African mesic grassland species. *Rangeland Ecology and Management* 62:572–578.

Giba, Z., D. Grubišić, S. Todorović, L. Sajc, Đ. Stojaković, and R. Konjević. 1998. Effect of nitric oxide-releasing compounds on phytochrome-controlled germination of Empress tree seeds. *Plant Growth Regulation* 26: 175–181.

Gilmour, C. A., R. K. Crowden, and A. Koutoulis. 2000. Heat shock, smoke and darkness: partner cues in promoting seed germination in *Epacris tasmanica* (Epacridaceae). *Australian Journal of Botany* 48(5): 603–609.

Gonzalez, S. L., and L. Ghermandi. 2012. Fire cue effects on seed germination of six species of northwestern Patagonian grasslands. *Natural Hazards and Earth Systems Science* 12: 2753–2758.

Gómez-González, S., A. Sierra-Almeida, and L. A. Cavieres. 2008. Does plant-derived smoke affect seed germination in dominant woody species of the Mediterranean matorral of central Chile? *Forest Ecology and Management* 255(5–6): 1510–1515.

Gómez-González, S., C. Torres-Diaz, and E. Gianoli. 2011. The effects of fi re-related cues on seed germination and viability of *Helenium aromaticum* (Hook.) H.L. Bailey (Asteraceae). *Gayana Botánica* 68(1): 86–88.

Hansen, A., J. S. Pate, and A. P. Hansen. 1991. Growth and reproductive performance of a seeder and a resprouter species of *Bossiaea* as a function of plant age after fire. *Annals of Botany* 67(6): 497–509.

Hidayati, S. N., J. L. Walck, D. J. Merritt, S. R. Turner, D. W. Turner, and K. W. Dixon. 2012. Sympatric species of *Hibbertia* (Dilleniaceae) vary in dormancy break and germination requirements: implications for classifying morphophysiological dormancy in Mediterranean biomes. *Annals of Botany* 109(6): 1111–1123.

Hong, E., and H. Kang. 2011. Effect of smoke and aspirin stimuli on the germination and growth of alfalfa and broccoli. *Electronic Journal of Environmental, Agricultural and Food Chemistry* 10(2): 1918–1926.

Humphrey, A. J., and M. H. Beale. 2006. Strigol: Biogenesis and physiological activity. *Phytochemistry* 67(7): 636–640.

Hutchings, A., A. H. Scott, G. Lewis, and A. B. Cunningham. 1996. *Zulu medicinal plants.* Pietermaritzburg, South Africa: University of Natal Press.

IUCN Red List of Threatened Plants. 2013. Compiled by the World Conservation Monitoring Centre. Retrieved 2013 from http://www.iucnredlist.org/.

International Plant Names Index, The. 2013. Retrieved 2013 from http://www.ipni.org.

Jain, N., G. D. Ascough, and J. Van Staden. 2008a. A smoke-derived butenolide alleviates $HgCl_2$ and $ZnCl_2$ inhibition of water uptake during germination and subsequent growth of tomato—Possible involvement of aquaporins. *Journal of Plant Physiology* 165(13): 1422–1427.

Jain, N., M. G. Kulkarni, and J. Van Staden. 2006. A butenolide isolated from smoke, can overcome the detrimental effects of extreme temperatures during tomato seed germination. *Plant Growth Regulation* 49: 263–267.

Jain, N., V. Soos, E. Balazs, and J. Van Staden. 2008b. Changes in cellular macromolecules (DNA, RNA and protein) during seed germination in tomato, following the use of a butenolide, isolated from plant-derived smoke. *Plant Growth Regulation* 54(2): 105–113.

Jain, N., W. A. Stirk, and J. Van Staden. 2008c. Cytokinin-and auxin-like activity of a butenolide isolated from plant-derived smoke. *South African Journal of Botany* 74(2): 327–331.

Jain, N., and J. Van Staden. 2006. A smoke-derived butenolide improves early growth of
tomato seedlings. *Plant Growth Regulation* 50: 139–148.

Jain, N., and J. Van Staden. 2007. The potential of the smoke-derived compound,
3-methyl-2*H*-furo[2,3-*c*]pyran-2-one, as a priming agent for tomato seeds. *Seed Science Research* 17: 175–181.

Jefferson, L. V., M. Pennacchio, K. Havens, D. Sollenberger, and J. Ault. 2007. *Ex situ* germination responses of midwestern USA prairie species to plant-derived smoke. *American Midland Naturalist* 159(1): 251–256.

Joel, D. M., J. C. Steffens, and D. E. Mathews. 1995. Germination of weedy root parasites. In J.
Kigel and G. Galili. 1995. *Seed development and germination.* New York: Marcel Dekker, Inc.

Jones, C. S., and W. H. Schlesinger. 1980. *Emmenanthe penduliflora* (Hydrophyllaceae): Further consideration of germination response. *Madroño* 27: 122–125.

Jovanović, V., Z. Giba, D. Djoković, S. Milosavljević, D. Grubišić, and R. Konjević. 2005.
Gibberellic acid nitrate stimulates germination of two species of light-requiring seeds
via the nitric oxide pathway. *Annals of New York Academy of Sciences* 1048: 476–481.

Jurado, E., M. Márques-Linares, and J. Flores. 2011. Effect of cold storage, heat, smoke and
charcoal in breaking seed dormancy *Arctostaphylos pungens* HBK (Ericaceae). Фyton
80: 101–105.

Jäger, A. K., A. Strydom, and J. van Staden. 1996. The effect of ethylene, octanoic acid and a
plant-derived smoke extract on the germination of light-sensitive lettuce seeds. *Plant
Growth Regulation* 19: 197–201.

Kandari, L. S., M. G. Kulkarni, and J. van Staden. 2011. Effect of nutrients and smoke solutions on seed germination and seedling growth of tropical soda apple (*Solanum
viarum*). *Weed Science* 59(4): 470–475.

Keeley, J. E. 1984. Factors affecting germination of chaparral seeds. *Bulletin of the Southern
California Academy of Sciences*: 113–120.

———. 1987. Role of fire in seed germination of woody taxa in California chaparral. *Ecology*
68: 434–443.

Keeley, J. E., and C. J. Fotheringham. 1997. Trace gas emissions and smoke-induced germination. *Science* 276: 1248–1250.

———. 1998a. Smoke-induced seed germination in California chaparral. *Ecology* 79(7):
2320–2336.

———. 1998b. Mechanism of smoke-induced seed germination in a post-fire chaparral
annual. *Journal of Ecology* 86: 27–36.

Keeley, J. E., and S. C. Keeley. 1987. Role of fire in the germination of chaparral herbs and
suffrutescents. *Madroño* 34: 240–249.

Keeley, J. E., T. W. McGinnis, and K. A. Bollens. 2005. Seed germination of Sierra Nevada
postfire chaparral species. *Madroño* 52(3): 175–181

Keeley, J. E., B. A. Morton, A. Pedrosa, and P. Trotter. 1985. Role of allelopathy, heat and
charred wood in the germination of chaparral herbs and suffrutescents. *Journal of
Ecology* 73: 445–458.

Keeley, S. C., and M. Pizzorno. 1986. Charred wood stimulated germination of two fire-
following herbs of the California chaparral and the role of hemicellulose. *American
Journal of Botany* 73(9): 1289–1297.

Keith, D. A. 1997. Combined effects of heat shock, smoke and darkness on germination of *Epacris stuartii* Stapf., an endangered fire-prone Australian shrub. *Oecologia* 112: 340–344.

Kempe, D. R. C. 1988. *Living underground: A history of cave and cliff dwelling*. London: Herbert Press.

Kenny, B. J. 2000. The influence of multiple fire-related germination cues in three Sydney *Grevillea* (Proteaceae) species. *Austral Ecology* 25: 664–669.

Krock, B., S. Schmidt, C. Hertweck, and I. T. Daldwin. 2002. Vegetation-derived abscisic acid and four terpenes enforce dormancy in seeds of the post-fire annual, *Nicotiana attenuata*. *Seed Science Research* 12: 239–252.

Kucera, B., M. A. Cohn, and G. Leubner-Metzger. 2005. Plant hormone interactions during seed dormancy release and germination. *Seed Science Research* 15: 281–307.

Kulkarni, M. G., S. G. Sparg, M. E. Light, and J. van Staden. 2006. Stimulation of rice (*Oryza sativa* L.) seedling vigour by smoke-water and butenolide. *Journal of Agronomy and Crop Science* 192(5): 395–398.

———. 2007a. Germination and post-germination response of *Acacia* seeds to smoke-water and butenolide, a smoke-derived compound. *Journal of Arid Environments* 69(1): 177–187.

Kulkarni, M. G., R. A. Street, and J. van Staden. 2007b. Germination and seedling growth requirements for propagation of *Dioscorea dregeana* (Kunth) Dur. and Schinz-A tuberous medicinal plant. *South African Journal of Botany* 73(1): 131–137.

Kępczyński, J., B. Białecka, M. E. Light, and J. van Staden. 2006. Regulation of *Avena fatua* seed germination by smoke solutions, gibberellin A$_3$ and ethylene. *Plant Growth Regulation* 49: 9–16.

Larkindale, J., J. D. Hall, M. R. Knight, and E. Vierling. 2005. Heat stress phenotypes of *Arabidopsis* mutants implicate multiple signaling pathways in the acquisition of thermotolerance. *Plant Physiology* 138: 882–897.

Light, M. E., B. V. Burger, D. Staerk, L. Kohout, and J. Van Staden. 2010. Butenolides from Plant-Derived Smoke: Natural Plant-Growth Regulators with Antagonistic Actions on Seed Germination. *Journal of Natural Products* 73(2): 267–269.

Light, M. E., M. I. Daws, and J. Van Staden. 2009. Smoke-derived butenolide: Towards understanding its biological effects. *South African Journal of Botany* 75: 1–7.

Light M. E., M. J. Gardner, A. K. Jäger, and J. van Staden. 2002. Dual regulation of seed germination by smoke solutions. *Journal of Plant Growth Regulation* 37: 135–141.

Lindon, H. L., and E. Menges. 2008. Effects of smoke on seed germination of twenty species of fire-prone habitats in Florida. *Castanea* 73(2): 106–110.

Lloyd, M. V., K. W. Dixon, and K. Sivasithamparam. 2000. Comparative effects of different smoke treatments on germination of Australian native plants. *Austral Ecology* 25(6): 610–615.

Long, R. L., J. C. Stevens, E. M. Griffiths, M. Adarnek, M. J. Gorecki, S. B. Powles, and D. J. Merritt. 2011. Seeds of Brassicaceae weeds have an inherent or inducible response to the germination stimulant karrikinolide. *Annals of Botany* 108(5): 933–944.

Ma, G., E. Bunn, K. Dixon, and G. R. Flematti. 2007. Comparative enhancement of germination and vigor in seed and somatic embryos by the smoke chemical 3-methyl-2*H*-furo[2,3-*c*]pyran-2-one in *Baloskion tetraphyllum* (Restionaceae). *In Vitro Cellular & Developmental Biology—Plant* 42(3): 305–308.

Maga, J. A. 1988. *Smoke in food processing*. Boca Raton: CRC Press.

Malabadi, R. B., N. T. Meti, G. S. Mulgund, K. Nataraja, and S. V. Kumar. 2012. Smoke saturated water promoted in vitro seed germination of an epiphytic orchid *Oberonia ensiformis* (Rees) Lindl. *Research in Plant Biology* 2(5): 32–40.

Mangnus, E. M. 1960. *Strigol analogues. Design, synthesis and biological activity*. PhD Thesis. Department of Organic Chemistry. NSR Center for Molecular Structure, Design and Synthesis. University of Nijmegen Toernooiveld. Nijmegen, The Netherlands.

Magnus, E. M., and B. Zwanenburg. 1992. Tentative molecular mechanism for germination stimulation of *Striga* and *Orobanche* seeds by strigol and its synthetic analogues. *Journal of Agriculture and Food Chemistry* 40: 1066–1070.

Martin, A. C. 1946. The comparative internal morphology of seeds. *American Midland Naturalist* 36: 513–660.

Merritt, D. J., M. Kristiansen, G. R. Flematti, S. R. Turner, E. L. Ghisalberti, R. D. Trengove, and K. W. Dixon. 2006. Effects of butenolide present in smoke on light-mediated germination of Australian Asteraceae. *Seed Science Research* 16(1): 29–35.

Merritt D. J., S. R. Turner, S. Clarke, and K. W. Dixon. 2007. Seed dormancy and germination stimulation syndromes for Australian temperate species. *Australian Journal of Botany* 55: 336–344.

Modi, A. T. 2002. Indigenous storage method enhances seedling vigour of traditional maize. *South African Journal of Science* 98: 138–139.

———. 2004. Short-term preservation of maize landrace seed and taro propagules using indigenous storage methods. *South African Journal of Science* 70: 16–23.

Moreira, B., J. Tormo, E. Estrelles, and J. G. Pausus. 2010. Disentangling the role of heat and smoke as germination cues in Mediterranean Basin flora. *Annals of Botany* 105(4): 627–635.

Morris, E. C. 2000. Germination response of seven east Australian Grevillea species (Proteaceae) to smoke, heat exposure and scarification. *Australian Journal of Botany* 48 (2): 179–189.

Morris, E. C., A. Tieu, and K. W. Dixon. 2000. Seed coat dormancy in two species of *Grevillea* (Proteaceae). *Annals of Botany* 86: 771–775.

Mulaudzi, R. B., M. G. Kulkarni, J. F. Finnie, and J. Van Staden. 2009. Optimizing seed germination and seedling vigour of *Alepidea amatymbica* and *Alepidea natalensis*. *Seed Science and Technology* 37(2): 527–533.

Nagase, R., M. Katayama, H. Mura, N. Matsuo, and Y. Tanabe. 2008. Synthesis of the seed germination stimulant 3-methyl-2H-furo[2,3-c]pyran-2-ones utilizing direct and regioselective Ti-crossed aldol addition. *Tetrahedron Letters* 49: 4509–4512.

Nelson, D. C., J. Riseborough, G. R. Flematti, J. Stevens, E. L. Ghisalberti, K. W. Dixon, and M. S. Smith. 2009. Karrikins discovered in smoke trigger *Arabidopsis* seed germination by a mechanism requiring gibberellic acid synthesis and light. *Plant Physiology* 149: 863–873.

Nelson, D. C., A. Scaffidi, E. A. Dun, M. T. Waters, G. R. Flemmatti, K. W. Dixon, C. A. Beveridge, E. L. Ghisalberti, and S. M. Smith. 2011. F-box protein MAX2 has dual roles in karrikin and strigolactone signaling in *Arabidopsis thaliana*. *Proceedings of the National Academy of Sciences* 108(21): 8899–8902.

Nikolaeva, M. G. 1969. *Physiology of deep dormancy in seeds*. Moscow: Izadetel'Stvo Nauka.

Norman, M. A., J. A. Plummer, J. M. Koch, and G. R. Mullins. 2006. Optimising smoke treatments for jarrah (*Eucalyptus marginata*) forest rehabilitation. *Australian Journal of Botany* 54(6): 571–581.

Parani, M., S. Rudrabhatla, R. Myers, H. Weirich, B. Smith, D. W. Leaman, and S. C. Goldman. 2004. Microarray analysis of nitric oxide responsive transcript in *Arabidopsis*. *Plant Biotechnology Journal* 2: 354–366.

Pate, J. S., R. H. Froend, B. J. Bowen, A. Hansen, and J. Kuo. 1990. Seedling growth and storage characteristics of mediterranean-type ecosystems of S.W. Australia. *Annals of Botany* 65(6): 585–601.

Penman, T. D., D. L. Binns, R. M. Allen, R. J. Shiels, and S. H. Plummer. 2008. Germination responses of a dry sclerophyll forest soil-stored seedbank to fire related cues. *Cunninghamia* 10(4): 547–555.

Pennacchio, M., L. V. Jefferson, and K. Havens. 2005. Smoke: Promoting germination of tallgrass prairie species. *Chicago Wilderness Journal: Best Practices in Conservation and Restoration* 3(3): 14–19.

———. 2007a. Where there is smoke, there is germination. *Illinois Steward* 16(3): 24–28.

———. 2007b. The inhibitory effects of plant-derived smoke on seed germination of *Arabidopsis thaliana*. *Research Letters in Ecology*. 1: 1–4.

———. 2010. *Uses and abuses of plant-derived smoke. Its ethnobotany as hallucinogen, perfume, incense and medicine*. New York: Oxford University Press.

Pepperman, A. B., and J. M. Bradow. 1988. Strigol analogues as germination regulators in weed and crop seeds. *Weed Science* 36(6): 719–725.

Pepperman, A. B., and H. G. Cutler. 1991. Plant-growth-inhibiting properties of some 5-alkoxy-3-methyl-2(5H)-furanones related to strigol. *A.C.S. Symposium Series* 443: 278–287.

Pierce, S. M., K. Elser, and R. M. Cowling. 1995. Smoke-induced germination of succulents (Mesembryanthemaceae) from fire-prone and fire-free habitats in South Africa. *Oecologia* 102: 520–522.

Pooley, E. 1993. *The complete field guide to trees of Natal, Zululand and Transkei*. Durban, South Africa: Natal Flora Publications and Natal Herbarium.

Preston, C. A., and I. T. Baldwin. 1999. Positive and negative signals regulate germination in the post-fire annual, *Nicotiana attenuata*. *Ecology* 80(2): 481–494.

Preston, A. A., R. Becker, and I. T. Baldwin. 2004. Is "no" news good news? Nitrogen oxides are not components of smoke that elicits germination in two smoke-stimulated species, *Nicotiana attenuata* and *Emmenanthe penduliflora*. *Seed Science Research* 14: 73–79.

Pyne, S. J. 2001. *Fire: A brief history*. Washington, DC: University of Washington Press.

Pérez-Fernández, M. A., and S. Rodríguez-Echeverría. 2003. Effect of smoke, charred wood, and nitrogenous compounds on seed germination of ten species from woodland in central-western Spain. *Journal of Chemical Ecology* 29(1): 237–251.

Queitsch, C., S. Hong, E. Vierling, and S. Lindquist. 2000. Heat shock protein 101 plays a crucial role in thermotolerance in *Arabidopsis*. *Plant Cell* 12: 479–492.

Rajjou, L., M. Duval, K. Gallardo, J. Catusse, J. Bally, C. Job, and D. Job. 2012. Seed germination and vigor. *Annual Review of Plant Biology* 63: 507–533.

Read, T. R., and S. M. Bellairs. 1999. Smoke affects the germination of native grasses of New South Wales. *Australian Journal of Botany* 47(4): 563–576.

Read, T. R., S. M. Bellairs, D. R. Mulligan, and D. Lamb. 2000. Smoke and heat effects on soil seed bank germination for the re-establishment of a native forest community in New South Wales. *Austral Ecology* 25: 48–57.

Reizelman, A., S. C. M. Wigchert, C. del-Bianco, and B. Zwanenburg. 2003. Synthesis and bioactivity of labelled germination stimulants for the isolation and identification of the strigolactone receptor. *Journal of Organic and Biomolecular Chemistry* 1: 950–959.

Reyes, O., and M. Casal. 2006. Seed germination of *Quercus robur, Q. pyrenaica* and *Q. ilex* and the effects of smoke, heat, ash and charcoal. *Annals of Forest Science.* 63: 205–212.

Riefner, R. E., Jr. 2009. *Plecostachys serpyllifolia* (Asteraceae) naturalized in California. *Phytologia.* 91(3): 542–565.

Roche, S., J. M. Koch, and K. W. Dixon. 1997a. Smoke enhanced seed germination for mine rehabilitation in the southwest of Western Australia. *Restoration Ecology* 5(3): 191–203.

Roche, S., K. W. Dixon, and J. S. Pate. 1997b. Seed ageing and smoke: Partner cues in the amelioration of seed dormancy in selected Australian native species. *Australian Journal of Botany* 45(5): 783–815.

Roeder, K., J. West, and B. Smith. 2011. Making a smoker to produce cellulose based smoke for treating forest tree seeds. *Native Plants Journal* 12(1): 27–29.

Rokich, D. P., K. W. Dixon, K. Sivasithamparam, and K. A. Meney. 2002. Smoke, mulch and seed broadcasting effects on woodland restoration in Western Australia. *Restoration Ecology* 10(2): 185–194.

Ross, S., and S. Rice. 2013. Germination of *Phacelia strictiflora* is enhanced by smoke. Poster Abstract, Botany 2013 Conference, New Orleans, July 27–31, 2013.

Ruthrof, K. X., M. C. Calver, B. Dell, and G. E. S. J. Hardy. 2011. Look before planting: Using smoke water as an inventory tool to predict the soil seed bank and inform ecological management and restoration. *Ecological Management and Restoration.* 12(2): 154–157.

Sato, D., A. A. Awad, Y. Takeuchi, and K. Yoneyama. 2005. Confirmation and quantification of strigolactones, germination stimulants for root parasitic plants *Striga* and *Orobanche*, produced by cotton. *Bioscience, Biotechnology and Biochemistry.* 69(1): 98–102.

Scaffidi, A., G. R. Flematti, D. C. Nelson, K. W. Dixon, S. M. Smith, and E. L. Ghisalberti. 2011. The synthesis and biological evaluation of labelled karrikinolides for the elucidation of the mode of action of the seed germination stimulant. *Tetrahedron* 67: 152–157.

Schwachtje, J., and I. T. Baldwin. 2004. Smoke exposure alters endogenous gibberellins and abscisic acid pools and gibeberellin sensitivity while eliciting germination in the post-fire annual, *Nicotiana attenuata. Seed Science Research* 14: 51–60.

Schwilk, D. W., and N. Zavala. 2012. Germination response of grassland species to plant-derived smoke. *Journal of Arid Environments.* 79: 111–115.

Shebitz, D. J., K. Ewing, and J. Gutierrez. 2009. Preliminary observations of using smoke-water to increase low-elevation Beargrass *Xerophyllum tenax* germination. *Native Plants Journal* 10(1): 13–20.

Singh, A., and P. Raizada. 2010. Seed germination of selected dry deciduous trees in response to smoke. *Journal of Tropical Forest Science* 22(4): 465–468.

Smith, T. M. 2006. *Seed priming and smoke water effects on germination and seed vigor of selected low-vigor forage legumes.* MSc Thesis, Virginia Polytechnic and State University, Blacksburg, VA. 102 pp.

Smith, M. A., D. T. Bell, and W.A. Loneragan. 1999. Comparative seed germination ecology of *Austrostipa compressa* and *Ehrharta calycina* (Poaceae) in aWestern Australian *Banksia* woodland. *Australian Journal of Ecology* 24(1): 35–42.

Sparg, S. G., M. G. Kulkarni, M. E. Light, and J. van Staden. 2005. Improving seedling vigour of indigenous medicinal plants with smoke. *Bioresource Technology* 96(12): 1323–1330.

Sparg, S. G., M. G. Kulkarni, and J. van Staden. 2006. Aerosol Smoke and Smoke-Water Stimulation of Seedling Vigor of a Commercial Maize Cultivar. *Crop Science* 46: 1336–1340

Stevens, J. C., D. J. Merritt, G. R. Flematti, E. L. Ghisalberti, and K. W. Dixon. 2007. Seed germination of agricultural weeds is promoted by the butenolide 3-methyl-2H-furo[2,3-c] pyran-2-one under laboratory and field conditions. *Plant and Soil*. 298(1–2): 113–124.

Strydom, A., A. K. Jäger, and J. van Staden. 1996. The effect of plant-derived smoke extract, N6-benzyladenine and gibberellic acid on the thermodormancy of lettuce seeds. *Plant Growth Regulation* 19: 97–100.

Sun, K., Y. Chen, T. Wagerle, D. Linnstaedt, M. Currie, P. Chmura, Y. Song, and M. Xu. 2008. Synthesis of butenolides as seed germination stimulants. *Tetrahedron Letters* 49: 2922–2925.

Tan, B. A. 2005. Smoke and mirrors in kangaroo paw germination. *Australian Plants* 23: 133–139.

Tang, Y., S. L. Boulter, and R. L. Kitching. 2003. Heat and smoke effects on the germination of seeds from soil seed banks across forest edges between subtropical rainforest and eucalypt forest at Lamington National Park, south-eastern Queensland, Australia. *Australian Journal of Botany* 51: 227–237.

Tavsanoglu, C. 2011. Fire-related cues (heat shock and smoke) and seed germination in a *Cistus creticus* population in southwestern Turkey. *Ekoloji* 20(79): 99–104.

Taylor, J. L. S., and J. van Staden. 1998. Plant-derived smoke solutions stimulate the growth of *Lycopersicon esculentum* roots in vitro. *Plant Growth Regulation* 26: 77–83.

Thomas, T. H., and I. Davies. 2002. Responses of dormant heather (*Calluna vulgaris*) seeds to light, temperature, chemical and advancement treatments. *Plant Growth Regulation* 37: 23–29.

Thomas, P. B., E. C. Morris, and T. D. Auld. 2003. Interactive effects of heat shock and smoke on germination of nine species forming soil seed banks within the Sydney region. *Austral Ecology* 28: 674–683.

———. 2007. Response surfaces for the combined effects of heat shock and smoke in germination of sixteen species forming soil seed banks in south-eastern Australia. *Austral Ecology* 32: 605–616.

Thomas, T. H., and J. van Staden. 1995. Dormancy break of celery (*Apium graveolens* L.) seeds by plant-derived smoke extract. *Plant Growth Regulation* 17: 195–198.

Thornton, M. A., T. H. Thomas, and N. C. B. Peters. 1999. The promotive effect of combustion products from plant vegetation on the release of seeds from dormancy. *Plant Growth Regulation* 28: 129–132.

Tieu, A., K. W. Dixon, K. A. Meney, and K. Sivasithamparam. 2001a. The interaction of heat and smoke in the release of seed dormancy in seven species from southwestern Western Australia. *Annals of Botany* 88: 259–265.

———. 2001b. Interaction of soil burial and smoke on germination patterns in seeds of selected Australian native plants. *Seed Science Research* 11: 69–76.

Tieu, A., J. A. Plummer, K. W. Dixon, K. Sivasithamparam, and I.M. Sieler. 1999. Germination of four species of native Western Australian plants using plant-derived smoke. *Australian Journal of Botany* 47(2): 207–219.

Tigabu, M., J. Fjellström, P. Odén, and D. Teletay. 2007. Germination of *Juniperus procera* seeds in response to stratification and smoke treatments and detection of insect-damaged seeds with VIS-NIR spectroscopy. *New Forests* 33(2): 155–169.

Todorović, S., Z. Giba, S. Živković, and D. Grubišić. 2005. Stimulation of empress tree seed germination by liquid smoke. *Plant Growth Regulation* 47: 141–148.

Tsuyuzaki, S. and Miyoshi, C. 2009. Effect of smoke, heat, darkness and cold stratification on seed germination of 40 species in a cool temperate zone in northern Japan. *Plant Biology* 11: 369–378.

USDA, ARS, National Genetic Resources Program. 2013. *Germplasm Resources Information Network—(GRIN)* [Online Database]. National Germplasm Resources Laboratory, Beltsville, MD. Retrieved 2013 from http://www.ars-grin.gov/cgi-bin/npgs/html/index.pl.

USDA NRCS. 2013. *Plants database* [Online Database] Natural Resources Conservation Service, Washington, DC. Retrieved 2013 from: http://plants.usda.gov/java/.

van Staden, J., A. K. Jäger, M. E. Light, and B. V. Burger. 2004. Isolation of the major germination cue from plant-derived smoke. *South African Journal of Botany* 70(4): 654–659.

van Staden, J., A. K. Jäger, and A. Strydom. 1995. Interaction between a plant-derived smoke extract, light and phytohormones on the germination of light-sensitive lettuce seeds. *Plant Growth Regulation* 17: 213–218.

van Staden, J., M. G. Kulkarni, G. D. Ascough, and M. Arruda. 2008. Smoke treatments improve growth and yield of commercially grown tomato and onion under greenhouse conditions. *South African Journal of Botany* 74(2): 381.

Vogl, R. J. 1974. Effects of fire on grasslands. In *Fire and Ecosystems: Physiological Ecology—A series of Monographs, Texts and Treaties.* Ed. T. T. Kozlowski, and C. E. Ahlgren. New York: Academic Press.

Wagley, C. 1957. *Santiago Chimaltenango; Estudio antropológico-social de una comunidad indígena de Huehuetenango.* Guatemala City, Guatemala: Seminario de Integracion Social.

Wang, Y. M., S. Q. Peng, Q. Zhou, M. W. Wang, C. H. Yan, H. Y. Yang, and G. Q. Wang. 2006. Depletion of intracellular glutathione mediates butenolide-induced cytotoxicity in HepG2 cells. *Toxicology Letters* 164: 231–238.

Wasser, C. H. 1982. Ecology and culture of selected species useful in revegetating disturbed lands in the West. FWS/OBS-82/56. Washington, DC: U.S. Department of the Interior, Fish and Wildlife Service, Office of Biological Services, Western Energy and Land Use Team. Available from NTIS. PB-83-167023.

Waters, E. R. 2003. Molecular adaptation and the origin of land plants. *Molecular Phylogenetics and Evolution* 29: 456–463.

Waters, E. R., G. Lee, and E. Vierling. 1996. Evolution, structure and function of the small heat shock proteins in plants. *Journal of Experimental Botany* 47: 325–338.

Waters, E. R., and B. A. Schaal. 1996. Heat shock induces a loss of rRNA-encoding DNA repeats in *Brassica nigra*. *Proceedings of the National Academy of Sciences, U.S.A.* 93: 1449–1452.

Waters, M. T., S. M. Smith, and D. C. Nelson. 2011. Smoke signals and seed dormancy. Where next MAX2? *Plant Signaling and Behavior* 6(9): 1418–1422.

Wehmeyer, N., L. D. Hernandez, R. R. Finkelstein, and E. Vierling. 1996. Synthesis of small heat-shock proteins is part of the developmental program of late seed maturation. *Plant Physiology* 112(2): 747–757.

Western Australian Herbarium. 2013. *FloraBase-The Western Australian Flora*. Department of Conservation and Land Management. Retrieved 2013 from http://florabase.calm. wa.gov.au/.

Wicklow, D. T. 1977. Germination response in *Emmenanthe penduliflora* (Hydrophyllaceae). *Ecology* 58: 201–205.

Wigchert S. C. M., and B. Zwanenburg. 1999. A critical account on the inception of *Striga* seed germination. *Journal of Agriculture and Food Chemistry* 47: 1320–1325.

Williams, P. R., R. A. Congdon, A. C. Grice, and P. J. Clarke. 2005. Germinable soil seed banks in a tropical savanna: seasonal dynamics and effects of fire. *Austral Ecology* 30(1): 79–90.

Willis, A. J., R. McKay, J. A. Vranjic, M. J. Kilby, and R. H. Groves. 2003. Comparative seed ecology of the endangered shrub, *Pimelea spicata* and a threatening weed, bridal creeper: Smoke, heat and other fire-related germination cues. *Ecological Management and Restoration* 4(1): 55–65.

Wills, T. J. and J. Read. 2002. Effects of heat and smoke on germination of soil-stored seed in a south-eastern Australian sand heathland. *Australian Journal of Botany* 50: 197–206.

Yamaguchi, S., M. W. Smith, R. G. S. Brown, Y. Kamiya, and T. Sun. 1998. Phytochrome regulation and differential expression of gibberellin 3β-hydroxylase genes in germinating *Arabidopsis* seeds. *The Plant Cell* 10: 2115–2126.

Zavala, N., and D. Schwilk. 2009. Smoke-induced seed germination in Texas grassland species. 94th Annual E.S.A. Meeting. Albuquerque, NM, August 2–7, 2009.

Zuloaga-Aguilar, S., O. Briones, and A. Orozco-Segovia. 2011. Seed germination of montane forest species in response to ash, smoke and heat shock in Mexico. *Acta Oecologica* 37: 256–262.

SUBJECT INDEX

SPECIES INDEX

INDEX OF COMMON NAMES

Pale Indian plantain, 88
Pale purple coneflower, 95
Pale rainbow, 136
Pale vanilla lily, 21
Palmer's penstemon, 252
Papaya, 14, 25
Paper daisy, 90
Paper heath, 232
Parrot bush, 219
Parry's blue-eyed Mary, 246
Paterson's curse, 114
Peach heath, 154
Pearl flower, 138
Pearly everlasting, 87
Peas, 163, 165, 170
Peegee hydrangea, 245
Pelargoniums, 174–5
Pennsylvania smartweed, 215
Penstemons, 247, 250–3
Pepper and salt, 243–4
Piligrass. *see* black speargrass
Pin heath, 140
Pincushions, 91, 228
Pine hill flannel bush, 260
Pine-leaved erica, 150
Pink beard heath, 138
Pink everlasting daisy, 108
Pink fountain triggerplant, 262
Pink heath, 142
Pink kunzea, 199
Pink mulla mulla, 79
Pink scholtzia, 201
Pink summer calytrix, 194
Pink watsonia, 43
Pink-flowered myrtle, 198
Pinks, 126
Pinweed, 129
Pipestems, 233
Pithy saw-sedge, 32
Pithy sword-sedge, 31
Pixie mops, 230
Plains coreopsis, 93
Platbaarvygie, 78
Plume rush, 65
Plume smokebush, 218
Plumed featherflower, 204
Pointed aneilema, 28
Pointed centrolepis, 26
Poison peach, 267
Pokeweed, 210
Poleo dorato, 189
Pollia, 28
Pom poms, 79
Poprosies, 260
Port Jackson heath, 144
Posy triggerplant, 263

Poverty brome, 52
Prairie acacia, 156
Prairie blazing star, 103
Prairie dropseed, 63
Prairie phacelia, 181
Prickly broom heath, 140
Prickly conostylis, 38
Prickly cryptantha, 114
Prickly ground-berry, 137
Prickly hakea, 224
Prickly roella, 124
Prince of Wales heath, 150
Procumbent yellow-sorrel, 208
Prostrate canary clover, 164
Proteas, 230–2
Puemo, 190
Purple clarkia, 206
Purple flag, 42
Purple prairie clover, 163
Purple tassels, 22
Purple vygie, 76
Purple wire grass, 48
Purple witchweed, 253
Pygmy patersonia, 42
Pyrenean oak, 172

Queen protea, 231
Queen triggerplant, 261
Queensland bluegrass, 54
Queensland brush box, 200
Quillay, 237

Rabo de gato, 189
Ragleaf bahia, 90
Rainbow pea, 165
Ramshorn, 224
Rapeseed, 115–16
Red crassula, 132
Red dead nettle, 183
Red fruited saw sedge, 31
Red grass, 50–1
Red gum, 197
Red hot poker, 25
Red ink sundew, 135
Red kangaroo paw, 38
Red lechenaultia, 176
Red pagoda, 229
Red root seed, 40
Red spider flower, 223
Red tassel flower, 96
Red velvet, 183
Redberry buckthorn, 235
Red-edged conebush, 226
Red-root cryptantha, 113
Reed triggerplant, 264
Reeds, 66–70, 72–3